Fat Chemistry
The Science behind Obesity

To Eleanor, Elias and Kate

Fat Chemistry
The Science behind Obesity

Claire S. Allardyce
Laboratoire de Chimie Organométallique et Médicinale,
Ecole Polytechnique Fédérale de Lausanne,
1015 Lausanne, Switzerland
E-mail: claireallardyce@netscape.net

RSCPublishing

ISBN: 978-1-84973-325-0

A catalogue record for this book is available from the British Library

Published by The Royal Society of Chemistry,
Thomas Graham House, Science Park, Milton Road,
Cambridge CB4 0WF, UK

Registered Charity Number 207890

For further information see our web site at www.rsc.org

Printed and bound in the United States of America

Preface

Certain professions have the habit of steering the topic of conversation in a standard direction: medical professionals guide the conversation towards a niggling health problem; plumbers remind people of a leaking tap. Chemists are faced with quite an unusual response: the conversation enters an awkward silence after words to the effect of "I didn't like chemistry at school." Nor did I. But it grew on me. And then I discovered that we can understand how the body works through chemistry and it became a bit of an obsession.

Back to the conversations; I found an area of chemistry that everyone was willing to discuss: fat chemistry. In particular, body fat chemistry. Surprisingly, slim people are just as interested in this as those that could do with shedding a few pounds. It seems the world is at war with fat and grateful for any ammunition they can pick up. I have collected loads. Some of these tips are much the same as what people have been saying for centuries: body weight regulation is a question of balance and a question of dose. Other pieces of information are relatively new, only discovered with recent advancements in science. One of these – the fact that body weight can alter the way we use our genes – was of sufficient concern to stimulate me to begin to write. This discovery dusts all the softness from obesity: it is now indisputably a deadly disease and our body weight has implications for future generations.

And so it is now time for conversations about fat chemistry to begin, because the only way to fight fat is together. To increase your knowledge of the enemy, read on. I would like to thank

Fat Chemistry: The Science behind Obesity
Claire S. Allardyce
© Claire S. Allardyce 2012
Published by the Royal Society of Chemistry, www.rsc.org

Jessica Adamson for many important conversations about fat and nutrition; Paul Dyson for many important conversations about chemistry; and the many people who have supported me in writing this book.

Contents

Fat Chemistry: The Science behind Obesity
Claire S. Allardyce
© Claire S. Allardyce 2012
Published by the Royal Society of Chemistry, www.rsc.org

CHAPTER 1

Why the Fuss about Obesity?

1.1 CLEARING THE NAME OF CHEMISTRY

Chemistry is a domain of science that often has negative con-
notations for the public. It is associated with poisoning, pollution,
destruction and devastation. And yet there is another side to this
science; a side on which the future of humanity depends.

If you take anything – plants, other animals or humans, the
bacteria or viruses that cause disease, the sea we bathe in, the air we
breathe or even the rock we stand on – and start chopping it down
into smaller and smaller pieces, eventually you will make a mixture
of atoms. Each type of atom is known as an element. Atoms are the
basic units of everything on Earth and in space. In most materials,
atoms are joined together to form molecules; few are poised
enough to go it alone. Helium is. It is one of just six noble gases;
elements so dignified that they reject liaisons with other atoms.
They mix, but they do not merge.

Helium atoms are found naturally in the mixture of atoms and
molecules we call air. True to type, they remain solitary and do not
join with other atoms to form molecules. The union process – a
chemical reaction – sometimes involves the exchange of heat:
chemical reactions may suck energy in and trap it in the chemical
bonds that link atoms together to form molecules, or they may
release energy formerly trapped in the same; sometimes in an

Fat Chemistry: The Science behind Obesity
Claire S. Allardyce
© Claire S. Allardyce 2012
Published by the Royal Society of Chemistry, www.rsc.org

explosive manner. Explosions are often based on the type of chemical reaction called combustion *i.e.* burning. In this type of
reaction, fuel molecules are cleaved and fused with oxygen to
release stored energy. This is the process that allows energy to be
released from food to power the body, but because of the potentially violent nature of combustion, it needs to be tightly constrained to support life.

Helium's lack of interest in chemical reactions makes it very
stable. It is reliable enough to give to children in balloons without a
repeat of the *Hindenburg* disaster, when a much larger balloon, an
airship to be exact, blew up dramatically back in 1937 killing many
of its crew and passengers. The disaster is famed because it was one
of the first tragedies to be caught on film and this footage is
probably the main reason why airship travel became rather
unpopular. Many more catastrophic events have been broadcast
since 1937, but this old film continues to be viewed by those
speculating on the cause of the blast. The favoured explanation is
that the balloon leaked. The gas in the airship was hydrogen.
Hydrogen is lighter than helium, therefore more effective at lifting
airships or balloons off the ground, but unlike helium it is an
explosive fuel; it loves liaisons. The explosion of the airship was
caused by the burning of hydrogen, which is an ideal fuel, because,
although this process releases plenty of energy, it produces no
unwanted gases or residues, only water – super-clean combustion.

The chemical nature of hydrogen and helium gives these gases
their similarities and differences: both gases are colourless,
odourless and allow balloons to float in air; yet one is stably celi-
bate and the other capriciously desperate for liaisons. This difference between the gases is determined by their electrons. Electrons
are one of three types of particle that make up atoms; the other two
are known as protons and neutrons. The protons are positively
charged and packed in the centre of the atom between the similarly
sized neutrons, which carry no charge at all. Together the protons
and neutrons form a dense core known as a nucleus. Compared
with protons and neutrons, electrons differ in size and space. In an
attempt to explain them in pictorial terms, imagine tiny midges
buzzing around a huge dung heap. The midges are the electrons
and the dung is the nucleus of the atom. In the image, the heap
needs to be at least 20 000 times bigger than an individual midge
because that is the estimated size difference between the electrons

and the nucleus of the smallest atom, hydrogen; it is estimated because electrons are really too small to measure. There are some differences between midges around a dung heap and atoms: for example, midges move in seemingly random ways, whereas the path of electrons follows a mathematical prediction called an orbital, and, whereas midges move alone, electrons prefer to be paired. Electrons are also charged, bearing a negative charge that neutralises the positive charge of the protons lodged in the nucleus of the atom. Take an isolated atom, which, outside the noble ranks of helium and its five kinsfolk, only exists in theory, and the fundamental unit would have the same number of protons and electrons. Whereas all the protons would be equivalent, the electrons would be found in two arrangements: those in stable, paired relationships with their own kind and those that are alone. It is the members of the singles club that are involved in bonds. Indeed, they are so desperate to make liaisons that they are seldom found alone in nature; rather they are given, received or shared between atoms to form chemical bonds and, fundamentally, it is the need to pair and share that sees few atoms going it alone. Helium's electrons are naturally paired – none is available to give or share – and so these atoms remain alone. In contrast, a lone hydrogen atom would have a single electron fidgeting and desperate to form a chemical bond. This desire is such that hydrogen gas is not made up of hydrogen atoms buzzing around on their own, but hydrogen molecules; two hydrogen atoms satisfy their needs by coming together so that the lonely electron in each atom can be paired and shared, and in so doing the hydrogen atoms form a molecule.

Like atoms, molecules are chemicals, but more complicated because they are made up of groups of atoms linked together. The smallest and arguably the simplest molecule is hydrogen gas. It is composed of the minimum number (just two) of the smallest type of atom (hydrogen). At the other end of the size scale are molecules so large that we can see them with the naked eye, for example strands of polymers, manmade fibres such as polythene or natural molecules, including our information storage molecule deoxyribose nucleic acid (DNA).

Chemistry is the science that strives to understand the behaviour of all chemicals, from the smallest to the largest, and their interactions with their surroundings. In light of this definition, it is possible to distil just about any part of science (and our lives) down to an essence of chemistry. However some powerful influences, in both science and marketing, dispute the dominance of chemistry. This opposition is mounted partly out of pride, claiming territory for other disciplines, but also because some factions want to put as much distance as possible between themselves and chemistry; chemistry has a certain reputation.

One particular industry that actively estranges chemistry is food manufacturing, demonstrated by the fact that marketing strategies often centre on listing groups of chemicals their products do not contain. Yet, despite the stand-off in public, chemistry and food manufacturing are closely intertwined; and have been so for many years.

Food science and chemistry grew up together. Virtually as soon as chemical know-how was developed, it was – and still is – translated into the domain of medicine and nutrition, improving the quality of our lives. Cooking is chemistry, despite what many chemists and cooks will have you believe. In the late 20th century, the close relationship between the food manufacturing and chemical industries came prominently into view, and the public did not like all that they found. A few decades before, the chemical industry had discovered some clever methods for processing, preserving and making food taste good. The methods generally involved adding extra ingredients; some were based on natural products and some were not. The food additives were (and are) tested for safety. For many tests the result was a "generally recognised as safe" (GRAS) categorisation. GRAS simply means that the chemical has no known detrimental health effects when used in the particular way evaluated, and in the particular quantities expected. For many food additives, the amounts in the final product are minuscule: tiny drops are added to vats of ingredients. At such low doses, even potential poisons can be generally recognised as safe. In this respect the GRAS categorisation is tried and tested: many natural poisons, such as cyanide, are found in the most natural of foods, but in such small amounts that they do not significantly affect health when the foods are incorporated into a balanced diet. Other chemicals that could poison the body in excess have health benefits in smaller doses. Take what we call vitamins as examples. In alphabetical order, vitamin A is the first: a fat soluble food component we need for good health, but an overdose, for example caused by one man eating an entire polar bear's liver, would be fatally toxic. There are many other examples of vital nutrients that become deadly poisons when consumed in excess, but generally there are few fatalities, because when they are part of the natural food source overconsumption is difficult. It is controlled by food availability, flavour and by our appetite. And with the food industry following such a tried and tested pathway, what could possibly go wrong? Perhaps too much success.

It is not clear whether the chemical industry ever imagined that their success in food production would be so sweeping – so revolutionary. But it was. And the consumers voted with their feet, selecting processed products over the fresh alternatives. Artificial flavours and flavour enhancers were central to this success. It is said that flavour is the reason why a particular food is bought repeatedly, and a few drops of the right chemical can make even the blandest products taste good. The cost of artificial flavours is often less than the packaging of the product. In addition, the most successful food products have been designed around innate preferences that optimise survival in the natural world: our partiality for salt, protein, fat and sugar. And so, with help from the chemical industry, some magic was woven in the factory kitchens and, for the first time, mass produced, low cost, convenient and tasty meal solutions became available. These products have enormous public appeal.

Within a decade convenience foods were incorporated into the daily diet of the majority of the population, in some cases to such an extent that certain food components were being consumed in excess of what had been anticipated. And health began to suffer. Nutritional deficiencies are one consequence of such dietary choices, but the odious links between chemical additives and health problems are more widely publicised; they are the consequences of excess. Towards the end of the 20th century, such links became headline material. History fuelled a rapid response. It was not the first time that the public had been shocked by food quality. Tampering with food began as soon as it was profitable, as soon as food production was outsourced beyond the family unit. But a lack of scientific understanding and a background of high levels of malnutrition and transmissible disease made the consequences difficult to trace. And so this first wave of dietary-induced illness festered for many centuries, until science became advanced enough to convict those who adulterated food. By that time – just outside living memory – special ingredients, from calves' brains to synthetic poisons, were being used to change the taste, texture or appearance of foods so that they fetched a higher price. When adulteration was at its height, few foods were untainted, from raw ingredients through to ready prepared products. Perhaps the most sickeningly ironic case was the use of toxic mineral dyes in lollipop treats for children. With hindsight it is clear that such practices significantly contributed to the poor growth and high rate of childhood death before the turn of the last century, when legislation was passed to protect the public. The public, logically, assumed the law would do what it set out to do; they were content and complacent. And so when the negative links between health and approved additives were revealed while the memory of the widespread and deadly incidents of food adulteration was still warm, much of society adopted a broad and bitter opinion of chemistry.

Decades on, the stigma associated with chemistry remains such that some manufacturers go to extreme lengths to distance themselves from this area of science; this can be so extreme it becomes ridiculous. For example, I was pleased that I had my camera with me on the day I purchased a coffee from a ready-to-eat food giant in the UK and found the following statement describing their organic milk plastered on the side of the cup: "it is natural, chemical-free and tastes good". This statement, if one will allow me the privilege of being pedantic, is meaningless: if their milk is free from chemicals then, technically, it is void of any atoms or molecules – a vacuum. Since molecules are what give our food flavour, how do they make a vacuum taste good, let alone put it in a paper cup? This enterprise has now modified its statement to say that it "creates handmade natural food avoiding the obscure chemicals", which again is not actually what I assume they mean to say. Recall the definition of chemistry: materials, both natural and synthetic, are all packed with chemicals. As a rule of thumb, synthetic chemicals are better characterised, more understood and less obscure than their natural cousins. Indeed, some of the current nutrition-linked health problems, including obesity, are not predominantly linked to synthetic food additives, but more to imbalances of natural chemicals – excesses and deficiencies of vital nutrients – often because of overconsumption of the same products that are rich in the synthetic additives.

Despite regularly having received a public flogging, just or unjust, chemistry has underpinned current nutrition and food safety. Only through advances in chemistry were cases of food adulteration proven such that legislation could be passed to protect the public from this type of poisoning. Only through chemical know-how is the food harvest preserved to prevent spoiling, protecting the public from some of the most deadly poisons known, including natural fungal toxins. Only through advances in chemistry have many nutritional deficiencies been identified and low-cost cures developed. Chemistry has been able to expand the food supply to provide an abundance to feed a growing population without ploughing up more natural habitat. And the role of chemistry in our lives does not stop there: this area of science is playing a principal part in determining the future of humanity. The current top four scientific challenges are energy, water, medicine and nutrition. Chemists are working in partnership with other scientists to tackle them all. This book focuses on the last two, although in truth they are all intertwined as increasing amounts of energy and water are being used to ensure that harvests are ample enough to feed well the growing population of the world.

1.2 CURRENT THREATS TO HUMAN HEALTH

In the domain of medicine, chemists are kept busy working on longstanding foes of humanity, and newcomers to the scene. If the truth be told, the human race currently faces major health scares that could wipe out a large percentage of the population if left unchecked. The World Health Organization (WHO) has been given the daunting responsibility of coordinating an integrated global alert and response system to contain what are considered to be the most prominent threats, should the need arise. On their hot list are the various forms of flu, severe acute respiratory syndrome (SARS), Ebola haemorrhagic fever, yellow fever, African trypanosomiasis, Crimean–Congo haemorrhagic fever, meningococcal disease, the return of the plague and so on. Other health threats currently challenging chemists include acquired immunodeficiency syndrome (AIDS) and the comeback of diseases formerly suppressed by vaccination, including tuberculosis and measles. In some cases a return is predicted to be accompanied by increased danger, due to drug resistance. Drug resistance is also causing new health threats, particularly in hospitals, where formerly controllable infections are becoming life threatening as they acquire defences against antibiotics; meticillin-resistant *Staphylococcus aureus* (MRSA) is one of the better known examples.

Some of WHO's hot-listed threats to human health come around in cycles, some are simmering away in remote parts of the world, while others just pop up out of the blue. An outbreak of any one of these diseases is likely to create fear amongst much of the population. Logically we need to be prepared for such events. To this end, scientists, including chemists, are busy trying to understand the mechanisms of initiation and progression of such diseases so as to develop new vaccines, treatments and containment programmes.

The spread of some diseases is restricted by their own characteristics with, for example, some infections being confined to the tropics simply because they are transported between victims by animals and insects that can only live in these warmer climates. Other diseases observe no geographical boundaries and can readily morph into epidemics. And others spread so rampantly that they scar many countries of the world and are awarded the prestigious "pandemic" status by WHO. Fortunately, medical knowledge is now sufficiently advanced to allow containment of most outbreaks of disease before they turn into pandemics, and often outbreaks do not even reach epidemic proportions because of intervention. Thus, although many diseases pose a real threat to human health, as long as we follow the guidelines provided by medical experts to protect ourselves, there is a good chance the threat of an epidemic will remain just a threat.

With such pressure on the world, one would be justified in asking the question: so why the big fuss about obesity? After all, isn't obesity just a natural consequence of gluttony and sloth? Such a cause suggests that self-control is the single cure and the health consequences of obesity are much like those of tobacco smoking – self-inflicted. And if the evolution of public opinion is any guideline,[1] in the future many ethical and political questions could be raised with respect to treating the obese, especially given the growing financial burden of the UK's excess body weight on the tax payer. Would this money be better spent tackling other medical issues?

One longstanding foe of humanity that appears frequently in the media threatening to become an epidemic is influenza, the viral infection more affectionately known as flu. The virus may have infected our hunter–gathering ancestors, but is most likely to have begun to blight humanity after the transition to farming; animal husbandry still facilitates the passage of new strains of the virus from birds and pigs to humans. The longstanding relationship between flu and humanity has given scientists many years to identify its cause, cures and vaccines. Given the nature of this disease, they have achieved this goal quite successfully. The malady normally strikes in the winter and confines some people to their beds, but most are up and about after a few days. And yet, amongst the weak and elderly, flu can be a killer. It is estimated that flu kills between one-quarter and half a million people worldwide each winter. Bouts of flu are even more serious when, occasionally, new strains of the virus emerge against which we have little immunity. Usually, the virus changes gradually, so that previous infections prime the immune system against re-infection, but a massive change can bypass immunological recognition and reduce the body's advantage in the fight against the disease. Such infections hit hard. These strains have the potential to sweep across the world and earn pandemic status whilst killing millions of people in a single season.

Each century there are about three flu pandemics. In the 20th century, the most serious outbreak was in the winter of 1918–1919; the disease was nicknamed the Spanish flu. The virus behind this pandemic is now thought to have originated in pigs rather than in Spain. The name was simply a result of misrepresentation as, then and now, most media coverage focuses on the plights of the rich and famous. Perhaps the richest and most famous sufferer in this outbreak was King Alfonso XIII, the then king of Spain. He was one of over 40% of the world's population infected by the virus, but not one of the estimated 40 million people who died. The victims were not just the young and the old (or even just the Spanish), but many people in the prime of their lives. Such devastation left a longstanding fear of flu. In 1957–1958, a strain of the avian influenza virus caused the so-called Asian flu pandemic, which resulted in one to four million deaths, depending on the source of the statistics. The 1968–1969 pandemic killed a number at the lower end of that range.[2] The most recent influenza pandemic involved swine flu, and it officially obtained this status between July 2009 and August 2010. As with the Spanish flu, all ages were affected; yet swine flu had a relatively mild impact on the annual death toll from influenza.

Long before swine flu was first detected, WHO had projected that a flu pandemic would arrive sooner or later. The regular

pattern of virus propagation allows an estimation of when major
bouts of disease are likely to break out and also their severity. It
was anticipated that the next bout would hospitalise two million
people and around half a million of them would die.[3] When flu
finally made it into the headlines, the story was not exactly as
predicted by WHO. Just like London buses, the serious outbreak
was late and then two came along together: avian flu and swine flu.
The avian flu virus lives primarily in birds, especially in Southeast
Asia, but by the end of 2009, over 400 cases of this flu strain had
been reported in humans, of whom nearly 60% died. The quantity
of deaths hardly altered the annual flu mortality statistics and does
not warrant epidemic, let alone pandemic, status; yet global
authorities were vigilant in containing its spread because of the
quality of its destruction: once infected, all individuals – young, old
or inbetweeners – declined very rapidly. Such severity stirred
community concern to implement containment programmes with
immediate effect, including the culling of many birds. These programmes undoubtedly reduced the impact of this outbreak on
humans and prevented a catastrophe.

In contrast to avian flu, swine flu did earn pandemic status. It
spread across 200 countries, infected half a million people and
caused 6000 deaths. Despite being a pandemic, swine flu did not
meet WHO's prediction, partly because the edge of its destruction
was softened by the containment programmes that worked on both
flu outbreaks simultaneously. The guidelines varied between
countries and also varied in inconvenience from relatively simple
precautions, such as increased hand washing, tiresome travel
restrictions and the humiliating use of masks in public places,
through to the more invasive participation in vaccination programmes. Inconvenience aside, it is clear that the awareness campaigns highlighted the seriousness of the threat and the public was
sufficiently concerned to participate and help reduce the spread of
the disease. We humans do not like being ill.

Currently, there is another longstanding foe of humanity on the
rampage that dwarfs the impact of flu on the population. This
more devastating disease was highlighted by WHO as an emerging
threat to human health back in 1990. By 1997, they announced it to
be an epidemic; in that year and every year since, despite the
issuance of tried and tested advice to prevent the spread, its incidence has only seemed to grow.[4] The number of people with this

disease has tripled in the UK during the last 20 years,[5] with the major escalation taking place in the last decade. Many people now describe its incidence as a pandemic, although such status is yet to receive official recognition. The disease does not care about its classification and the magnitude of its destruction continues to grow. And the people it kills are not the frail. Whereas most deaths, including those that result from the regular rounds of flu, occur in people over the age of 70, this other disease culls people of all ages. In the UK, it is named, under various different codes, as a cause on one-quarter of death certificates,[6] and while this number is rising, it is still thought to be an underestimate. In the USA it has been suggested to be responsible for as much as half of the premature death toll.[7–9] The discrepancy between the recorded and the actual statistics arises from two sources: first, the symptoms often melt away in the run up to death and, second, there is a reluctance to assign this label to people, even after their demise: the disease is heavily stigmatised. Various taboos do not allow us to mention it, even when the symptoms obviously affect our friends and family.

The difficulty in estimating the prevalence is just one example of the elusiveness of this disease; its growing incidence is another. There are guidelines and recommendations to contain this epidemic, and these are largely the same guidelines and recommendations that have protected humans successfully throughout history, but they no longer seem to have much effect. More than a decade after it was awarded epidemic status, it continues to challenge chemists, exasperate medics, frustrate politicians and provide a goldmine for quacks; and the number of people affected continues to grow. The extent of this scourge is so great that the same number of people that WHO predicted would die prematurely worldwide from the formerly impending flu epidemic will die in the UK and USA alone from this other malady. Few people would argue that such a disease deserves the attention of top scientists – including chemists – around the world, until we mention its name: this all-powerful disease and obesity are one and the same.

1.3   DIAGNOSING OBESITY

The diagnosis of obesity has one of two foundations: body weight and fat chemistry. The first type of diagnosis is much simpler and more widely used. The diagnostic chart goes under the name of the body mass index (BMI), which was derived from formulas first proposed by Lambert Adolphe Jacques Quetelet in the 19th century. He was a Belgian maths obsessive who revolutionised the social sciences by introducing statistics in an attempt to improve public health;[10] and then he turned his attention to body fat. Links between body weight and health were established long before Quetelet's time, but his work turned a loose association into a clear statistical correlation. Still, his charts did not become a medical standard until the middle of the 20th century, when science was sound enough to elucidate the molecular details of excess body weight and the far-reaching consequences; and costs.

Although the dearest price of obesity is human misery (physical and emotional pain, reduced quality of life and reduced life expectancy), the most easily measured cost is financial. Apart from direct treatments for the body fat, many obese people need to be treated for diabetes, circulatory problems and other maladies caused by unhealthy fat chemistry. These treatments come from a limited healthcare budget. Waiting lists in hospitals are growing and the cost of healthcare is rising for one and all. Back in 2006, the annual direct health costs of the health problems linked to excess body weight in the UK were estimated to be one billion pounds.[11,12] It was predicted that this cost would continue to rise, increasing three-fold by 2010. That number was an underestimate: the UK Department of Health has released figures that suggest that, in 2010 alone, the direct cost of unhealthily high bodyweight to the National Health Service was over four billion pounds. Healthcare is just one of the financial burdens the public faces as a growing portion of the population become obese. Infrastructure, from buildings through to buses, has to be adapted to accommodate bigger people, all at a cost. Support for the outsized ranges from larger and reinforced seats in public places, home aids to compensate for disability, and financial supplements to support those with reduced working capacity. Additionally, many governments around the world are investing proactively in education and public awareness programmes to encourage the population to stay trim.

It was the financial cost of obesity that led to the smoothing of Quetelet's formula in the 1950s, to create the BMI as we know it. This index is presented with four zones: underweight, healthy, overweight and obese. Statistically, the length and quality of life, along with one's expected health, alter according to the zone. Being in the normal zone is, of course, healthiest. To work out your BMI, your body weight in kilograms is divided by your height in metres squared. If the number that results from your equation falls between 19 and $22\,kg\,m^{-2}$ then, statistically, you are in the healthiest group; up to a value of $25\,kg\,m^{-2}$ is considered to be OK. If your value is less than $19\,kg\,m^{-2}$ you are likely to be underweight. This is the group with statistically the lowest life expectancy. The prediction is biased because many terminal illnesses are accompanied by weight loss, pushing their victim into the underweight bracket by the time of death. Similarly, health suffers when the body is overweight; a BMI value of 25 to $29.9\,kg\,m^{-2}$. Anyone who generates a number of over $30\,kg\,m^{-2}$ is classed as obese and, if your body weight enters this zone because of fat accumulation, then your health is severely suffering.

The obesity epidemic has not only manifested itself by pushing an increasing sector of the population into the obese zone: once the $30\,kg\,m^{-2}$ barrier has been crossed, the pushing power intensifies and the big become bigger and bigger. As a result, two extra zones have been tagged recently onto the high end of the BMI: morbidly obese and super obese. The morbidly obese zone spans BMI values between 35 and $39.9\,kg\,m^{-2}$; the super obese zone categorises anyone with statistics over that figure. With each step up there is a significant health decline.

Based on the current BMI zones, the number of obese people in the world is around half a billion. More than half of the world's population live in countries where too much fat kills more people than too little and, averaged globally, at least 10% of the population is obese;[13] at least 10% of the world's population has a reduced quality of life because of excess body fat. However, these figures are flawed because body weight is not a failsafe method of diagnosing obesity. On one hand, a lean, muscular athlete may fall into the BMI obese zone. On the other hand, resting below the BMI cut-off for obesity does not mean that your fat chemistry is on safe ground. It is easy to forget that this disease is not just a question of excess body weight, but it involves changes in body chemistry triggered by fat. People with a certain BMI fall into a statistical bracket in which the altered body chemistry associated with obesity is almost certain – as is the health decline – but for some people this change occurs with much smaller fat deposits than average. These people are chemically obese, although their BMI suggests that they are just overweight.

A better indication of fat-related health may be obtained using the waist-to-hip ratio (WHR). Unlike the BMI, the WHR aims to gauge the quality of fat storage and not just the quantity; it is also based on simpler maths. The fat stored around the midriff is more than just an energy reserve; it is a hormone-producing gland. Small deposits are important for fertility; larger deposits generate more than a healthy level of hormones. These hormones cause declining health and a tendency to gain weight. The WHR aims to gauge just how much fat is pumping out hormones. The calculation simply involves dividing your waste measurement by that of your hips and as long as both measurements use the same units, it does not matter whether you choose metric or imperial. In general, a ratio of 0.7 for women and 0.9 for men is considered healthy, with men having a higher allowance because of their naturally thicker waists. When the WHR rises just 0.1 points above these values there is a noticeable health decline; a health decline that is the result of chemical obesity.

Still, the WHR has its weaknesses, of which genes are associated with one: some people have a genetic predisposition to store fat disproportionately on the buttocks, hips and thighs, giving a healthy WHR measurement even if the belly fat has been deposited above the healthy mark. So it seems that the only failsafe way of diagnosing obesity is to screen the body and the blood for signs of the detrimental changes in chemistry that occur as a consequence of excess adiposity.

When fat chemistry is factored into the equation, the number of obese people in the world is estimated to double: one in five people worldwide have fat deposits that are significantly damaging their health; in some countries nine out of ten people are obese. These statistics are why, for the first time in human history, obesity is classed along with influenza and other potentially fatal diseases as a major medical challenge. Obesity is a disease that cannot be ignored.

1.4 THE DEEP ROOTS OF PREJUDICE

Throughout history, obesity has been considered quite distinct from a bit too much body fat. Weight gain, like grey hair, affects most people as they get older and to different extents. Some succumb gracefully, others fight both signs of ageing with whatever means are available. Those that allow themselves to grow too fat with age are often stereotyped as "characters". Just think of an authentic period drama based a century or more ago and in it you will probably find a stout king, a podgy cook or a well-padded wet-nurse. These stereotypes demonstrate the presence of excess body fat among different ranks of society, but in the majority of cases these people were plump, not obese. Obesity was rare and different enough from being overweight to earn its own categorisation – and its own reputation.

The Ancient Greeks were amongst the first peoples to document being excessively fat as a medical problem,[14] but the name for the condition has Latin roots. Hippocrates, the father of modern medicine, called obesity "corpulence", a term derived from the Latin word for the body, *corpus*. He understood that overly full bodies create health problems, and the nearest translation of his original words reads: "corpulence is not only a disease itself, but the harbinger of others."[15] The other diseases that follow the onset of obesity are the result of changes in body chemistry attributable to unhealthy fat stores, and these problems can now be mapped at a molecular level; molecular markers show when the excessive fat stores are damaging a person's health.[16]

Centuries after Hippocrates had described corpulence, the term "obesity" became favoured and by 1620 it appeared in print in Thomas Venner's *Via Recta* to describe the occasional gentleman or gentlewoman whose health was suffering because of over-indulgence. This term was derived from the Latin word *obesus*, which means stout, plump and vulgar; in this way the label carries a sense of moral judgement. And the remedy – simply balancing diet, sleep and other factors, as suggested by Hippocrates centuries before – loaded the burden of breaking out of their vulgar disposition directly onto the shoulders of the patient. Most were successful; a few were not.

Perhaps in a drive towards political correctness, or perhaps to return to Hippocratic roots, the term "corpulence" replaced "obesity" through the 18th and 19th centuries; only for the fashion to revert in the 20th century. Irrespective of the name, throughout human history the disease, its incidence and its remedy remained more or less the same, with the emphasis on treatment being the patient's own responsibility.

Modern science shows that, ironically, the perturbation in body chemistry caused by unhealthy fat reserves makes it harder to lose weight.[17] However, before science could map the molecular details of this relationship, the information was common knowledge. Back in 1916, Anna M. Galbraith wrote, in her book *Personal Hygiene and Physical Training for Women*, "Any weight above two and one-half pounds to the inch in stature may be considered as excessive, inasmuch as it adds nothing to one's mental or physical efficiency, and is frequently the forerunner to obesity."[18] This demonstrated the knowledge that a certain amount of unhealthy body fat, in this case crossing into the current overweight BMI zone, promotes further weight gain. Now this understanding has scientific backing, the amount of prejudice against the obese may be reduced, but there is an enormous barrier to break down because for most of history they were mocked – unless they could protect themselves by their wealth. In the 5th century BC, Aristophanes wrote the Ancient Greek comedy *Plutus* in which he described the obese as, to put it bluntly, stupid gluttons, figures of mockery and disgust: "bloated, gross and preseniled... they are fat rogues with big bellies and dropsical legs, whose toes by the gout are tormented." The stereotype is, in part, founded on statistics. The diseases triggered by obesity include oedema (dropsy) and gout. In addition, obesity is itself a symptom of disease. Throughout most of history, these diseases were usually of the genetic type – health problems rooted in the basic units of heredity – that alter body chemistry to promote unhealthy levels of body fat. Alström's syndrome is an example in which heredity dictates symptoms that go beyond fat metabolism to include mental retardation. Bardet–Biedl's syndrome similarly couples obesity with a low intellect. These syndromes result from systemic developmental problems that affect numerous organs and glands to produce broad symptoms. Such powerful diseases are – and always have been – rare, but when other types of obesity were also rare, they were responsible for a greater percentage of the total cases of excess body fat and, thus, the fattest people in the population were often preseniled; the exceptions were kings, lords and their families.

Wealth protects against prejudice and discrimination. When food is scarce and obesity rare, the novelty factor makes body fat the ultimate sign of wealth, one that demonstrates visually how a person can afford to indulge, and an investment that cannot be stolen. Blub is certainly a must-have among some peoples of Africa. In some environments, fat may increase the chance of survival: when the food supply is unreliable, gluttony ensures that sufficient reserves are in place in case of future famines, and laziness preserves these reserves. For much of human history the fluctuations in the food supply are speculated to have been so serious that they kept obesity at bay. The possibility of becoming overweight was confined to a few persons who were determined enough to fight for more than their share, be their motivation driven by genetic disease or human nature.

Times have changed. Fat is no longer a must-have in most nations and the principal cause is no longer inheritance (either genetic or wealth), but seemingly a choice of diet and lifestyle. With the wealth factor gone, public opinion of obesity is largely based on Aristophanes's ideas: there are people who believe that if the obese are not stupid gluttons, they are slothful gluttons. The prejudice comes with moral judgement so powerful that these roads to obesity are two among the list of the top Seven Deadly Sins. The view is strengthened by an inner understanding: greed and laziness are part of human nature and we have all experienced the pull towards a high calorie indulgence or taking the car when we could easily walk. It seems logical to believe that these sinful tendencies can cause excessive body weight gain if left unchecked; especially now that chemistry's interference with food technology has increased the supply and opened obesity to one and all. Yet some people escape the snare. They exert sufficient restraint to counterbalance their primitive instincts and stay slim; trim has become vogue. Such is the desirability of their prize that they could easily become smug and condescending towards those with seemingly less control. After all, everyone makes their own choices, and if one chooses to overeat one should suffer the consequences. Prejudice grows from this apparent inequality.

Once snared, fear of descending further into the body fat trap is vividly illustrated by the lengths people will go to try to protect themselves. When willpower cannot be trusted, jaw wiring or stomach banding are just some of the techniques used to reduce food consumption. Internal bands are now offered to squeeze the stomach invisibly, but the external banding practice began centuries before. Take Britain, for example, over the 400 years from the 16th century, when Henry VIII began his reign, until the beginning of the 20th century, when Queen Victoria ended hers. This was a time when high-born women (and wannabe high-born women) conformed to social pressures by squeezing into bone-deforming corsetry while poorer, less conforming country folk chose to spread a little and respire freely.

Victorian social pressures completely reversed the wealth–weight relationship in England and the standing of the Commonwealth had a global effect on fashion. But the trend began many years before, near the timeline transition from BC to AD. The Ancient Greeks, spanning the 9th to 3rd centuries BC, preferred a fuller figure: for men full muscular development and for women full because of fat. The women depicted in their statues were of the curvy type and, in general, skinny waifs were pitied because popular opinion assumed that they could only be skinny because of poverty and not by choice. At this time, the wealthy were weighty. Across the water in Rome a few centuries on, the transition was in full swing. The gradually growing food supply opened obesity to the masses and so the wealthy visibly demonstrated their superiority by staying slim; they were mostly successful despite larger than necessary food supplies. Desirable Roman women were stick-thin, a historical preference recorded most colourfully in the work of a comic. Terence was his name. He wrote a play entitled *Eunuchus* that was very popular in its time. The plot includes a character who is in love with an old-fashioned girl with voluptuous curves. In her defence, he belittles the current vogue for the super-slim: "mothers strive to make them have sloping shoulders, a squeezed chest so that they look slim... Though she is well endowed by nature, this treatment makes her as thin as a bulrush. And men love them for that!" These young women were possibly underweight, which is a state that brings with it a number of health problems, but the description implies that women bound their chests, much like the Victorians bound their waists, to appear thinner than they naturally were. The play touches on the suffering these young women experienced in order to be fashionable and, more worryingly, reveals the part their parents played in its infliction. Thin was a valuable asset then; and it still is now.

For centuries, stories of the rich and famous and their body weight have intrigued people: both their struggles in fighting fat and the remedies they tried. In some cases, successfully taming the drive towards obesity has led to fame and following. Fast-forward to the modern day and celebrity status is obtained by those who fight fat successfully. The wealth–weight relationship is fully overturned in much of the world; rather than wealth, poverty is now associated with overeating. For the most part, the elite are able to demonstrate their superiority by defying the call of nature and controlling their gluttony; perhaps channelling their urges into forms of greed that sustain their wealth as oppose to eating it away. When they struggle, they can hire the expertise of nutritionists and personal trainers, following Hippocrates' guidance on weight regulation and balancing their energy intake with expenditure to stay trim. They can seemingly buy their way out of obesity, but for some people no amount of money will buy freedom.

1.5 A GENETICALLY PROGRAMMED DISEASE

Throughout history there have been some strong-willed, wealthy, big people who would rather have been slim, because there is a cause of obesity stronger than greed and laziness: genetically predestined disease. Such examples of obesity include pleiotropic syndromes, such as Alström's and Bardet–Biedl's, which were the likely inspiration behind Aristophanes's "fat fools" – characters used to add a certain type of humour to his plays – through to cases where increased body fat is the only apparent symptom, until the health decline it harbingers sets in. Daniel Lambert is one of the earliest cases in the latter category to be sufficiently documented to demonstrate that he was neither stupid, slothful nor a glutton.

Lambert was an 18th-century jailer in the Leicester County Bridewell detention centre, until he became too big to work – he weighed 335 kg (over 52 stone) before his death in 1809. His BMI was $103 \, \mathrm{kg \, m^{-2}}$.

In the early 1800s people with a BMI of $30 \, \mathrm{kg \, m^{-2}}$ or more – the obese – were rare. A BMI of $103 \, \mathrm{kg \, m^{-2}}$ was rarer still, so rare that in Lambert's lifetime he was one of a kind. At this time, thin was in vogue and although some men of wealth were plump (women were usually bound by corsetry), perhaps even pushing into the obese category, most actively strove to lose weight and were successful with the remedies of the time – balancing diet with exercise – so that those in the higher BMI zones were responding to an innate calling, and for this illness they were mocked. Lambert outgrew the heftiest of them. Lambert was over five times heavier than a healthy modern man. The *Oxford Dictionary of National Biography* entered his name alongside the phrase "the most corpulent man of his time in

England" – perhaps adopting Hippocrates' naming scheme was considered less judgemental. His fame travelled across the waters in the form of a wax model that went on to tour the USA after his death. Lambert's extraordinary size has remained an international curiosity right up to the modern day, and none are more interested in his tale than his townsfolk. In the town where Lambert once worked, current curiosity maintains a permanent museum exhibition dedicated to the man, along with several pubs and restaurants boasting his name.

Growing up a stone's throw from where Lambert spent most of his life, I have no excuse for not having sat in his custom-made chair simultaneously with several others. The design of chairs could be altered, albeit at a cost, but the town planners' budgets were more limited; narrow streets with narrower doorways were inconvenient for a cumbersome bloke. Cleverly, Lambert turned inconvenience to an advantage; a lucrative advantage. He challenged locals and visitors to a wagered race and then used his bulk to block their passage in the narrowest streets so that he came in first. These games are just one example of Lambert's fitness, despite the fat. They are also just one example of his enterprising ingenuity; although he was obese, Lambert was neither slothful nor stupid.

Lambert enjoyed races because he was a sporting fellow. However, as his weight increased so did the challenge. A point came when many activities became impractical and then impossible; not through a lack of fitness, but through the sheer inconvenience of his form. He had to turn sideways to enter some buildings, and in others the door width simply barred him. He was outsized.

When the cash from the races dried up, he headed off in his custom-made carriage to where the streets are paved with gold – London. Once there, poverty forced him to find creative and extreme ways of making money: he charged punters a shilling to see his phenomenal fat rolls. And then he squandered his revenue on the fees of the top physicians he consulted about his fat; consultations with the sole purpose of making it go away. The spectacle reeks of desperation. Size reduced his career options to one: an exhibit.

Stories written about Lambert during his lifetime reveal a compassionate person: he helped many of his wards during their stay at the Bridewell, which, at the time, was quite unheard of because these people were being detained for one of the worst crimes of that era – they were poor. The *Leicester Journal*, the leading local newspaper of the time, described Lambert as a "happy fellow" – fat and happy, like Father Christmas. Despite being obese, Santa has achieved immortality. Lambert was not so lucky. He died suddenly at the age of 39, possibly from a pulmonary embolism.[19]

Lambert's tombstone and obituaries give a snapshot of public opinion at the time: fond tributes usually include references to what was then a remarkable size. In life, Lambert shied from the scales, as though embarrassed by the truth, but in death his weight was engraved in stone next to his name; his friends remembered his size as part of his personal achievement. However, the intrigue was not always based on admiration. Sometimes it was based on scorn and disgust; he was weird enough to pay a shilling to see – both in the flesh and replicated in wax. The story of the spectacle of the most corpulent 19th-century Englishman continues to intrigue. Lambert's story has been retold many times. One of the earliest examples is in *The Eccentric Mirror*, a series of books that catalogued male and female characters renowned for peculiar characteristics. More recently, for example, Jan Bondeson awarded Lambert a personal chapter in his 21st-century book about medical marvels,[20] only to retell the story six years later under the encompassing categorisation of a "freak".[19] The public response suggests that Lambert had crossed the barrier of an acceptable physique. He was neither a king nor a lord and so wealth did not protect him from the prejudice that categorised fat as the most obvious part of the stupid glutton stereotype. Lambert challenged this view. His mental and physical agility undermined aspects of the longstanding prejudice against the obese. Lambert was not your typical fat man of the time.

After a fit and active early life, Lambert was obese by the age of 23. He tried to diet during his lifetime, but, as reported by the *Leicester Journal*, he still increased in "bulk". This led to a frustration with which the modern obese individual can empathise, a frustration that can be amplified by the fact that outsiders may put this lack of success down to cheating. Lambert does not escape this type of scorn. Some modern-day medics have given Lambert the label of primary obesity: caused by gluttony and sloth. Yet the historical accounts of his life do not tick all the boxes necessary for this diagnosis. First, there are claims that he ate modestly. Those who label Lambert as a glutton explain away these claims as lies: dieters deceive themselves because of their greed. But what about the many accounts of his fitness made by people at the time of his reign as the fattest man alive? Next to almost every chronicle of his girth is an entry regarding his sporting achievement, physical fitness and agility. Here was a man who may have lied about his food intake, but he could not have lied about his ability to walk long distances at whim. His vast body weight meant that a leisurely stroll would have consumed the calories of an intensive workout for the average man. Lambert's actions are those of a man who burnt enough calories to stay trim on a normal diet – perhaps even an excessive diet – and yet he was huge. And Lambert appeared to have all the right ideas when it came to weight control: increasing physical activity, abstaining from alcohol, and restricting his consumption of food to single servings. What is more, he had significant motivation for weight control: he was literally outsized. He was excluded from public transport and public places because he could not squeeze through the doors. His clothes and custom-made furnishings were expensive and there were no public funds to cushion these costs. His body weight drove him to social isolation and financial ruin. Enough incentive to lose weight, but his body size still increased. And finally, there is no reason for Lambert to have denied gluttony: the public were intrigued by his fat and they did not care how it came to be deposited. If Lambert had been a glutton, it would have made no difference to his followers. Such a confession may have even had a financial advantage: think of the entertainment value of enticing the fattest man alive with food! Indeed, why wage a race through the town when he could have better satisfied his greed (and laziness) by challenging punters to an eating competition? These ideas and more are why some experts who have studied Lambert's life, including Philip French, the current curator at the Newarke Houses Museum in Leicester, England, defend the idea that Lambert had a medical complaint and was not a large eater.[21] In his home town, Lambert's fans believe that his obesity was not a result of the sinful tendencies of gluttony and sloth, but was caused by his fat chemistry, chemistry that resulted from the complement of genes he had inherited.

1.6   MODERN LAMBERTS

While his size made him famous for over two centuries, there have only been a few self-confessed Lambert wannabes. Fear of fat runs too deep. Like kings and lords, sumo wrestlers (*rikishi*) rise above the scorn by wearing specific robes to distinguish their status. They try hard to follow Lambert's pattern of fit and fat. But it is a tough ride, a dedicated way of life with the advantage of cultural respect and the disadvantage of a life-expectancy 10 years shorter than average. To gain such weight, *rikishi* follow a disciplined schedule, including skipping breakfast and then eating a hearty lunch before taking a postprandial nap to allow the fat to flow straight to their stores. Yet even with such dedication, they generally have a BMI in the 30s and 40s, occasionally pushing into the 50s, only rarely do they become heavier, and they have yet to enter Lambert's zone. Hawaiian-born Konishiki Yasokichi was the heaviest *rikishi* ever in sumo. His BMI crossed into the 80s. He used his weight to his advantage, but despite all efforts, he was beaten in the final stages of becoming the first foreign-born sumo grand champion. And he was beaten in the fat polls by people who were trying to be slim.

Carol Yager is thought to top the all-time fat polls. This Michigan lass had a BMI of nearly $260 \, \text{kg m}^{-2}$ at her peak – two and a half times that of Lambert. She died in 1994 at the tender age of 34 because of obesity-related complications. Another North American, Jon Brower-Minnoch, was the heaviest man recorded in history. He peaked at an enormous $635 \, \text{kg}$, boasting a BMI of $191 \, \text{kg m}^{-2}$. Obesity killed Brower-Minnoch at much the same age as it killed Lambert. Both Yager and Brower-Minnoch became obese without much effort; their weight gain began in childhood and such an early start traditionally suggests a genetically hard-wired cause, an inheritance that perturbs body chemistry to such an extent that the fat rolls suffocate their victims before they experience the mid-life crisis. Therefore these people were not Lambert wannabes, but Lambert gottabes; they wanted to be trim, but their genes dictated a different fate.

Some supersized people claim to be in control; they claim to be Lambert wannabes. After Brower-Minnoch's death, the Mexican Manuel Uribe entered the 2007 Guinness World Records as the fattest man alive, weighing eight-fold more than the average healthy man. He himself blamed his vastness on diet and life-style and to prove his point, after weighing in for the world record, he began work on a second for the most rapid weight loss. Rapid weight loss amongst the supersized is hardly surprising: morbidly obese people suffer from oedema (fluid retention), which can be shed quickly. The fluid is also a significant contributor towards their premature deaths. Uribe believes he suffers from primary obesity, the politically correct term for unhealthy fat deposits caused by gluttony and sloth. However, despite actively trying to lose weight he has yet to enter the healthy BMI zone. If his weight gain was simply a publicity stunt, the cost to his health is likely to have been greater than any financial return. Galbraith's warnings to women apply just as much to men.

In 2011, an Arizona mum Susanne Eman began a mission to top Lambert's size. After years of struggling to keep her body weight in the healthy zone, she flipped her objectives in the opposite direction: she embarked on a mission to become the biggest woman in the world. Her quest, she believes, is benevolent: to help the supersized come to terms with their weight. A doctor describes her game as Russian roulette. Indeed, although her health has yet to decline, it will. Obesity has become the number one preventable cause of premature death, and such "achievement" is diluted by the number of fat people in the world. Lambert-sized chairs are now mass-produced and the big are getting bigger. Despite the occasional rise to fame, the number of unhealthily fat people today abolishes any prestige associated with being obese. The prestige has gone, but the prejudice remains. Even after Lambert's efforts, the obese are still considered to be lazy, greedy, uneducated and even of low intelligence.

1.7   THE QUANDARY OF THE EPIDEMIC

With such baggage, could the obesity epidemic simply be due to greed and laziness? This million-dollar question can be answered by taking a look at the trillion-dollar weight loss industry: a massive number of people are actively investing in remedies supplied by an industry that is built on how wretched excess body weight makes one feel. The feelings stem from social pressures and prejudices, but also the reduced quality of life and poor health that accompany excessive fat storage. And the focal point is self-judgement: diet products are purchased by the individual seeking to lose weight, not by their friends or family – or even their enemies – trying to make a point.

Yet buying into the weight loss industry is no protection against unhealthy body weight gain, as the extent of the worldwide epidemic demonstrates. As this industry grows, body weight seems to be spiralling out of control. The extent of the problem is evidence that, despite various claims, there is no miracle for weight loss. Some anti-obesity drugs are being trialled, but success is usually dependent on a lifestyle overhaul – and tolerating the side-effects. There are no slimming pills that negate the need for effort and will-power, and there are no surgical methods that are entirely successful unless coupled to changes in the diet. After reading this book, I hope it will be clear that the problem of fat extends beyond the blubber we see and that pharmacological intervention that solely targets fat deposition may simply treat the symptom without con-sidering the cause; these remedies may be cosmetic rather than health-restoring. Still, these products remain on the menu because of consumer demand. The demand becomes even more incompre-hensible when considered in the light of the existence of the well known, freely available, simple, successful and scientifically proven recipe for weight loss and health-restoration: eating less and doing more. This cure for obesity is based on the theory of energy balance.

To compare energy intake versus expenditure, and thereby pre-dict body weight change, imagine an old-fashioned set of shop scales, like the one on the cover of this book. In the days of yore, the shopkeeper mounted metal blocks of different weights on one side and produce on the other to give their customers the right measure. Now, let us adapt that image for energy balance. The energy balance weighs calories and not mass. On one side we mount the food we eat and the balance moves according to the calories it contains. On the other side we mount metal blocks weighted according to the calories we burn. If the food calories are greater than the calories used the balance is in favour of weight gain and if the calories used are greater than the food calories then we lose weight. Simple.

Understanding the energy balance has allowed the majority of our ancestors to successfully regulate their body weight in a healthy zone. The wide availability of this trouble-free, low cost obesity cure seems to suggest that people can choose whether they are going to be one of the thousands to die pre-maturely from fat or whether they want to go on living by balancing diet and lifestyle. This recipe for weight loss has been used effectively and to a dangerous extent for millennia, particularly by women. Terence's play *Eunuchus*, first performed in 161 BC, is clear evidence of its successful application: mothers are said to have controlled their daughters' body weight such that, "If one is a little plumper, they say she is a boxer and they reduce her diet." The story was much the same in recent history: super-slim was iconic in the 1960s and 1970s, and although the icons themselves may not have chosen the healthiest methods of weight control, throughout the general population most people stayed trim by balancing their diet and activity levels (rather than resorting to the use of narcotics or the surgeon's scalpel).

The desire to be slim remains in today's culture and, in keeping with the historical pattern, some people manage to control their weight using these methods; others do not. Yet there is a significant difference in the statistics: throughout most of history, inheritance – genetically predestined disease or wealth – was the main reason for fat, causing obesity for a maximum of one in twenty people; today, amongst the fattest nations of the world the figure is close to being reversed: one in twenty people have a healthy body weight and the remainder are fatally fat.

The obesity epidemic has gripped the world, as more remedies are launched. The tried and tested cure – the energy balance – no longer seems to work. This state of affairs suggests that modern obesity does not follow the same rules as that longstanding foe of humanity that has the same symptoms.

In times gone by, obesity was usually the sign of a genetic disease. Today, this same name is an umbrella term that groups together any disease in which fat is destroying health, and increasingly the cause is environmental. Despite being preventable and curable, the environmental cause is equally as powerful as its genetically programmed twin, certainly more powerful than the human desire to stay slim. Hospitals and forced weight loss regimes that focus on the fat fail in these circumstances; just as they did for Lambert.

Today, modern medicine and public resources soften the individual blow by treating the diseases that fat causes, and cushioning the financial implications of immobility and poor health such that the obese can expect to survive beyond Lambert's 39 years, but the quality of life is not ideal and the cure remains as elusive for us as it did for him; or so it seems.

1.8 FIGHTING FOR A FUTURE

The current patterns in the incidence of obesity challenge the traditional understanding of the disease. Therefore, it is time we dismissed our prejudices and preconceived opinions and took another look at unhealthy body weight – a molecular look – to find the origin of the modern epidemic. To find a new direction that will bring this epidemic to an end, we need to understand how the body works to be able to judge how food affects health and weight. Some of this type of data is available now; some will be unveiled in the future. We need to move away from referring to excess fat as natural, as the body is designed to avoid it. We need to understand the importance of being vigilant with our body weight because this is central to good health.

Throughout history human folk have battled with body weight and, unless a genetic disease slung the game, people won. Today, we have the same distribution of genetic variation, more knowledge, better nutrition and more free time to exercise. Yet despite knowing the importance of healthy body weight, an increasing sector of the world's population is troubled by Lambert-style obesity. The current accumulation of scientific data suggests that the problem is probably a consequence of diet and lifestyle, not primary obesity, which is considered to be driven by an element of choice, but a forced effect caused by imbalances in fat chemistry that drive appetite beyond needs and reduce the metabolic rate. Science has now shown that the environment can trigger changes in fat chemistry that mimic the force of genetically predestined obesity – even working through the genes. The profile of this disease includes a drive to eat that goes beyond the body's needs and a level of complexity that defies the traditional remedies.

Under the spotlight of chemistry, the molecular details of obesity demonstrate a way out. They suggest that the current epidemic is a nutritional problem. Throughout history, nutrition has caused waves of disease – epidemics of illness caused by what we eat. Traditionally, nutritional problems were caused by a lack of food in general, and so such blights were accompanied by weight loss; weight gain was considered to be a sign of healthy nutrition. Today, this link no longer stands and malnutrition can be hidden under a blanket of fat. Interspersed among the nutritional deficiencies that result from an imbalanced diet are examples of a new dietary problem caused by excess: over-nutrition. Over-nutrition with respect to energy is publicised as a cause of obesity, but other excesses are misunderstood as harmless, even considered healthy. Still, they affect body chemistry and in some cases the effect is to promote fat storage.

At the beginning of the 16th century, the Swiss Renaissance physician, botanist, alchemist, astrologer and general occultist Paracelsus made a statement that should be engraved on our minds: the right dose differentiates a poison and a remedy. His actual words were somewhat more elaborate "*Alle Ding' sind Gift, und nichts ohn' Gift; allein die Dosis macht, daß ein Ding kein Gift ist*", which loosely translates to: "all things are poison, and nothing is without poison, only the dose makes something not a poison". According to this philosophy, the obese are being poisoned by their fat because they have exceeded the healthy dose of energy. Yet delving into the molecular details of obesity suggests that fat is not the only poison in their systems. Vitamin and mineral excesses can be poisonous too. So can food additives. Health can suffer greatly with too much of a good thing. There is increasing evidence to suggest that the good things we take in excess to cause obesity are not the obvious calorie-containing food components, but virtually calorie-free chemicals in our foods; chemicals that meddle with our appetite and metabolism. Perhaps chemistry should take some of the blame for obesity; certainly this area of science has contributed to setting the stage for the disease. However, the public also has to accept responsibility because nutritional advice is freely available and printed on many products. The obesity-promoting excesses and deficiencies are characterised and the information is already in the public domain. A sector of the population has accepted this responsibility and remains trim. Several institutions are responding to the findings. For example, at the Interdisciplinary Obesity Centre, Kantonsspital St Gallen, Switzerland, the medical team led by Bernd Schultes performs a broad nutritional analysis on their patients before prescribing treatment; their findings show that many of the obese are deficient in several nutrients that are important for body weight control.[22] Today, fat can be considered as much a cause of one disease as a symptom of another: modern malnutrition.

Modern malnutrition is a mixture of deficiencies and excesses that meddle with our fat chemistry and affect our physical and mental well-being; obesity is the most noticeable symptom, but there are others that affect mental and physical health, threatening future epidemics. Soberingly, for the first time in history, obesity has slipped out of the mid-life crisis to plague children. Being young and fat is linked with a higher chance of disability in adulthood and also of premature death.[13] There are other consequences for the next generation: new evidence is emerging to suggest that the fate of our children is affected by our own body weight. Overweight parents produce children with innate metabolic disturbances that promote unhealthy body weight.[23,24] Therefore, the new trend of obesity among children could create a vicious cycle of declining health for generations to come. The extent of the problem is such that if action is not taken now, the children born today are not likely to enjoy lives of the same high quality and length as their parents.[25]

REFERENCES

1. C. Scott-Clark, in *Sunday Express*, BMA backs doctors who refuse to treat smokers, 1994.
2. E. J. Walker and R. G. Webster, *Am. Sci.*, 2003, **91**, 122–129.
3. *WHO Report on Global Surveillance of Epidemic-prone Infectious Diseases*, World Health Organization, Geneva, 2000.
4. *Obesity: Preventing and Managing the global Epidemic*, World Health Organization, Geneva, 2000.
5. P. M. L. Skidmore and J. W. G. Yarnell, *QJM*, 2004, **97**, 817–825.
6. M. Duncan, M. Griffith, H. Rutter and M. J. Goldacre, *Eur. J. Public Health*, 2010, **20**, 671–675.
7. A. D. Lopez, C. D. Mathers, M. Ezzati, D. T. Jamison and C. J. L. Murray, *Lancet*, 2006, **367**, 1747–1757.
8. G. Sankaranarayanan, J. D. Adair, T. Halic, M. A. Gromski, Z. Lu, W. Ahn, D. B. Jones and S. De, *Surg. Endosc.-ULTRAS*, **25**, 1012–1018.
9. D. Yeste, J. Vendrell, R. Tomasini, L. L. Gallart, M. Clemente, I. Simon, M. Albisu, M. Gussinye, L. Audi and A. Carrascosa, *Horm. Res. Paediatr.*, 2010, **73**, 335–340.
10. G. Eknoyan, *Nephrol. Dial. Transplant.*, 2008, **23**, 47–51.

11. *Tackling Obesity in England (House of Commons Papers)*, Stationery Office, Great Britain, 2001.
12. UK National Audit Office, *Tackling Child Obesity – First Steps: Hc 801, Session 2005–2006: Report by the Comptroller and Auditor General*, NAO, Great Britain, 2006.
13. World Health Organization, *Fact sheet 311: Obesity and overweight*, http://www.who.int/mediacentre/factsheets/fs311/en/index.html, accessed November, 2011.
14. D. Haslam, *Obes. Rev.*, 2007, **8**, 31–36.
15. D. W. Haslam and W. P. T. James, *Lancet*, 2005, **366**, 1197–1209.
16. H. M. Stanciola Serrano, G. Q. Carvalho, P. F. Pereira, M. d. C. Gouveia Peluzio, S. d. C. Castro Franceschini and S. E. Priore, *Arq. Bras. Cardiol.*, **95**, 464–472.
17. J. K. Sethi and G. S. Hotamisligil, *Semin. Cell Devel. Biol.*, 1999, **10**, 19–29.
18. A. M. Galbraith, *Personal Hygiene and Physical Training for Women*, W. B. Saunders Co., Philadelphia, 1916.
19. J. Bondeson, *Freaks: The Pig-Faced Lady of Manchester Square & Other Medical Marvels*, Tempus Publishing, Stroud, UK, 2006.
20. J. Bondeson, *The Two-headed Boy, and other Medical Marvels*, Cornell University Press, New York, 2000.
21. P. French, Newalke Houses Museums, Leicester, UK, personal correspondence, 2011.
22. B. Ernst, M. Thurnheer, S. M. Schmid and B. Schultes, *Obesity Surg.*, 2009, **19**, 66–73.
23. S.-F. Ng, R. C. Y. Lin, D. R. Laybutt, R. Barres, J. A. Owens and M. J. Morris, *Nature*, 2010, **467**, 963–103.
24. M. Wills-Karp, J. Luyimbazi, X. Xu, B. Schofield, T. Y. Neben, C. L. Karp and D. D. Donaldson, *Science*, 1998, **282**, 2258–2261.
25. S. J. Olshansky, D. J. Passaro, R. C. Hershow, J. Layden, B. A. Carnes, J. Brody, L. Hayflick, R. N. Butler, D. B. Allison and D. S. Ludwig, *N. Engl. J. Med.*, 2005, **352**, 1138–1145.

CHAPTER 2

It all Began with Change. . .

So far in human history there have been three broad stages of
nutritional development. The first was the hunter–gatherer era,
when humans lived hand to mouth from Nature's resources. In
this period there were nutritional problems, mainly caused by
food shortages. In the second stage, farming began. Our ancestors had more control over the food supply and the efficiency of
food production increased. Nutrition began to change. Science
was not sound enough to offer reliable guidance and so our
ancestors followed their instincts, favouring fatty, sweet, salty,
sour and protein-rich meals, and the easiest route to them. New
nutritional problems began to surface, this time due to both deficiencies and excesses. Despite the increased food supply, life
expectancy was close to that of the Stone Age; and it remained
there right through until the third stage of nutritional history
was established. This last stage was more objective. It was in this
period that science began to tame Stone Age partiality. And it
was successful. The epidemics of diet-induced diseases – from
scurvy through to heart disease – were significantly suppressed
within decades of discovering the derivation of the problem. The
foundations of human nutrition that we use today were laid.
However, the nutritional needs then were different from those
present now, science and technology were comparatively primitive and no one understood the molecular details of fat chemistry. Obesity is the result of these oversights.

Fat Chemistry: The Science behind Obesity
Claire S. Allardyce
© Claire S. Allardyce 2012
Published by the Royal Society of Chemistry, www.rsc.org

2.1   SHATTERING THE STONE AGE DREAM

The Stone Age spans the time from when people started to use stone tools, and encompasses copper metallurgy up to the point when bronze was introduced. The artefacts suggest that early Stone Agers were hunter–gatherers, but by the end of this period most were farmers.

Modern views of the Stone Age often betray a romantic infatuation with the hunter–gatherer lifestyle. One well publicised fan of hunter–gathering is Tom Varley. He commented, "It's great to live like this... you can do what you want, you could sleep all day and you had no job to go to".[1] His "Stone Age experience" was part of a project organised in 1997 by Science Wonder Productions, who charged visitors to observe Varley as he worked in his 3 acre enclosure in Austria. Varley built himself a house made of mud and sticks, and for five months "turned back the clocks" to life in the Stone Age. He spent most of his day collecting wild foods, fishing, tending to the small garden he had planted and occasionally hunting. He wore skins and furs to keep warm during the day, but each evening hung up his hide-cloak and went back to the 20th century to summarise his experiences on the company's web site.

Although people who spend a few weeks hunter–gathering generally enjoy the experience, the reality of this lifestyle was – and still is – dramatically different from their short experiences. The picture Varley painted of a lazy life, free of the stresses of modern employment, was also free of most of the stresses that living in the Stone Age involved. Daily life-threatening challenges for a Stone Ager included the constant threat of attack (from animals or fellow humans) and unseen dangers, such as parasitic worms, food poisoning, frostbite and malnutrition. The overall consequence of the Stone Age hardships was that, although they had the maximum number of children feasible, population growth was slow; it follows that the premature death rate must have been high.

In September 1991, walkers discovered a body in the Ötztal Alps, in Italy. It was so well preserved it was presumed to be recent; until clumsy excavation revealed Stone Age equipment. The man, who became known as Ötzi, is estimated to have died over 5000 years ago. A Stone Ager with skin caused a wave of excitement throughout the world and, inflicting only minimal damage, scientists set to work to find out as much as they could about the life of the man who became Europe's oldest natural human mummy. At the time of death it is estimated that he was about 1.65 m tall, 50 kg in weight and about 45 years of age. He lived close to where he died, but he had been travelling. His intestine contained two meals, both of game, grain, roots and fruits. The grain was highly processed einkorn, possibly in the form of bread. The food combination suggests that his settlement mixed agriculture with hunting and, because this cereal is harvested in late summer and his death was in spring, food storage mechanisms were in place. The first meal had been consumed in a conifer forest, but wheat and legume pollen were mixed with that from the pines, suggesting that domesticated crops may have been planted nearby; perhaps he lived in a clearing in a forest where farming could take place. Carbon tattoos matched appropriate acupuncture or acupressure points to treat the osteochodrosis and spondylosis indicated by his bones; medical advances had been made. Beau's lines on his nails were tell-tale signs of three serious bouts of illness in the past six months, giving an indication of medical limitations. Traces of poisonous metals in his hair, suggesting exposure to arsenic during metal working, and blackened lungs, caused by long periods sheltering in smoky rooms, demonstrated that Stone Age health suffered because of pollution. His bone density and shape suggested malnutrition during his childhood, perhaps due to lack of food availability, but also because of the parasites that were hitching a ride in his gut and blocking the absorption of nutrients from his food. His bone shape also suggested that he walked long distances across hilly terrain; he may have been a shepherd or, given the used weapons in his possession, a boundary guard.

The snapshot Ötzi gives us into prehistoric life is richly detailed, but singular. Commonly, such images are more sketchy yet more numerous. Take, for example, the shell mounds, which are thought to be amongst the earliest human necropoles. The structures were not built carefully from carved stone, but constructed from randomly piled shells and organic matter. Nevertheless they are impressive because the waste was placed over such long periods of time that the pile became more of a hill. One of the earliest archaeological excavations of such a site in the United States was of the now destroyed Emeryville shell mound, which consisted of five or six mounds made by the Native Americans who lived along the mouth of the perennial Temescal Creek, on the east shore of

San Francisco Bay.[2] Over what is estimated to be a 3000-year period, they piled up their kitchen waste and animal remains, along with their dead, to create prime compost. When discovered, the mounds were reported to be over 60 feet (18 m) high and some 350 feet (110 m) in diameter – until they were levelled to make way for industry, amusement parks and so on.

The Californian mounds are not unique, but part of a worldwide prehistoric cultural practice. On the west coast of the USA, examples of shell mounds have been preserved and include those in the Everglades National Park. Jump south and find the Brazilian equivalents, called *Sambaquis*, dotted along the south and south-east coasts. Just like those in North America, the mounds include kitchen waste, artefacts and tombs. Switch continents to Africa and well known shell mounds include those in the Banc d'Arguin National Park, nestled between the Sahara desert and the Atlantic Ocean, and those in the west in the Saloum Delta National Park, Senegal. Go down under to Australia and visit Fraser Island, just off the coast of Queensland. It is the largest sand island in the world; it also has shell mounds. In South Korea the metropolitan city of Busan has an exhibition hall, the Dongsam-dong, named after the shell mound from which its exhibits were salvaged. Here the finds have been interpreted to suggest a lifestyle.

The Dongsam-dong shell mound exhibition halls display many kinds of fish, shells and animal bones, as well as patterned pottery, earthenware, necklaces, bracelets, earrings, and other ornaments made out of shells, bones, teeth, stone, clay and jade. The different types of pottery demonstrate the presence of trade links between Korea and Japan, certainly by the end of the Stone Age (the Neolithic period). The tools suggest that the Dongsam-dong people depended on hunting and gathering, tapping into coastal and terrestrial resources, while practising small-scale cultivation of grain. The artefacts also suggest a sophisticated lifestyle with artistic preferences and spiritual beliefs; jewellery has been suggested to be as much for decoration as a talisman. These pieces were engraved with ritualistic symbols of the natural elements that so heavily influenced the people's lives.

Along with the artefacts skeletons have been found. The way the skeletons' former owners were laid to rest gives an indication of their culture. In South Korea, bones are found in small pits alongside burial goods such as pottery, stone axes, projectile points, shell bracelets and ornaments. Children were buried in jars. In other mounds, the bodies were painted with red ochre before being buried. Such practices suggest respect for the dead and possibly a preparation for an afterlife.

From bone alone we can get snapshots of life before death. Skeletons may just seem like a bunch of bones used to frighten little kids at Halloween, but to a trained eye they can tell a whole story: the overall shape of the bones can age the owner; thickening, weakening, wear and tear give an insight into lifestyle; chips, dents and bends can reveal a bit of personal history; radio-dating can place the bone's owner in a period of time; and high tech analyses with dauntingly complicated names, such as laser ablation inductively coupled mass spectrometry (LA-ICP-MS), alongside a simple tape measure, can gauge nutrition and food availability. With all these techniques to hand, the evidence suggests that Stone Age life was tough; certainly tougher than Varley's experience.

In the late 19th century a German anatomist and surgeon named Julius Wolff noted that different people's bones, although similar, were not the same, and the differences were linked to lifestyle. From his observations, he made a generalisation that is yet to be disproved, and this longstanding acceptance within the scientific community has promoted the idea from the status of theory to a law. Wolff's law states that living bone will adapt according to the loads it experiences. It describes how movement and work strains bones such that the bits strained the most are thickened and strengthened. The remodelling is initiated in the internal architecture of the softer, spongy trabeculae, followed by changes in the more compact external cortical bone portion.[3] The law also holds in reverse: as loading decreases, bones become weaker. And so the relative strength of bones – even fossilised ones – shows how people live; life stories are written in bone. For example, probing the relative thickness of the femur (thigh bone) and humerus (upper arm bone) can determine whether the deceased exercised more on foot, perhaps due to a nomadic lifestyle, or used arm strength for practices such as the extensive grain grinding required for a cereal-based diet. Stone Age foragers living in the Alps inhabited a rugged terrestrial environment that resulted in thickening of their leg bones relative to their arm bones, yet Andamanese Islanders who spent much of their time paddling canoes had relatively thick arm bones. Lifestyle can also give an insight into diet: canoe use could suggest that the people feasted on more marine foods than their terrestrially based brothers and sisters. But why speculate when inferences can be confirmed with chemical facts: remember that dauntingly named technique abbreviated to LA-ICP-MS? It uses lasers to vaporise tiny amounts of material, chopping them into the smallest pieces – atoms – which are fed into the main body of the machine to determine the atomic composition of the mix and produce a fingerprint-like pattern. By analysing the atomic content of bones, we can distil the details of our ancestors' diet. For example, this type of study suggests that in South Korea the pre-historic people buried in the shell mounds depended on game rather than marine foods and that their children were weaned before the age of 18 months.[4]

Changes in bone composition within the population over time can give an insight into dietary flux and how it affected health. In a study of about 250 human remains, from the San Francisco Bay shell mounds, the skulls were found to decrease in size and the incidence of malformations, like those left by rickets, rose as farming was introduced to the region; the researchers attributed the problem to calcium deficiencies.[5] Similarly, amongst the remains of people found in the Great Valley of central California, skeletal defects can be linked to dietary change – or rather, dietary strain. When the population density was low – the glory days of hunter–gathering – the people feasted on fish, meat and vegetables without the need to wear out their teeth on grain-grinding. To allow the population to grow, the food supply may have needed to be expanded, with the introduction of new flavours. The new addi-tions to the menu in the Great Valley indicate that the choice was forced rather than preferred: prehistoric people began to supple-ment their preferred menu of hunted prey with acorn-meal, and as they did so malnutrition became the norm.[6]

Acorns are not a farmed crop – oak trees take too long to grow from seed to be viable in agriculture – but Mother Nature planted an abundance of these trees in Neolithic California and the nuts provided a similarly abundant source of carbohydrate. There are various types of oak growing along the valley – Tan Oak, Black Oak, California Live Oak, Valley Oak and so on – and the nuts fall to the ground plentifully each autumn, ready to be scooped up. Conveniently, the worm-ridden ones can be easily spotted as they tend to lose their little hats – the extra weight of the worm coupled to its wiggling breaks the nut away from the cap and the tree. Still, after selection and collection, the nuts are hardly ready to eat; they are laced with poison. The first step in processing is drying, either with or without their protective coating. Next, all the dried acorns need to be shelled before grinding. The final step towards making a meal is leaching – an essential process to remove the toxins the trees put in their seeds to stop them being eaten. The leaching process involves extensive washing and periodic tasting to determine when the bitter toxins have gone. When the acorn-meal is palatable, it can be incorporated into stews and breads, as the recipes require.

Given the fact that we humans generally favour protein foods and we are intrinsically lazy creatures, opting for the easiest solution to a meal, and given the effort involved in preparing acorns to produce a less favourable energy source than the fish and meat that once plentifully nourished the prehistoric Californians, it is logical to conclude that the gradually increased consumption of acorns occurred because hunger overcame both their laziness and fussy food preferences. There are a number of possible reasons for such hunger, but the overwhelming likelihood is that the human population outgrew the favourable food supply for hunter–gathering. Hunter–gatherers need a high-fat, high-protein diet to balance their active lifestyle and keep them healthy, or just about healthy. When farming was in its infancy in the Fertile Crescent of California, the population density was low, allowing the people to benefit from the best diet Mother Nature could offer, including birds, fish and calcium-rich molluscs. However, over sufficiently long periods of time, even the slowest population growth can accumulate and exceed the supply of good food. It is likely that by the time the practice of farming had spread around the world to reach what is now San Francisco, the population was in great need; need sufficient to cause bone deformities attributable to malnutrition. Survival in such an environment necessitated that the people ate all they could, rested when they could, and opted for the most energy-dense food available; an apparent recipe for obesity when food supplies grow beyond needs.

2.2 MODERN HUNTER–GATHERERS

Nutrition and medicine are thought to be the two main factors that curbed population growth amongst hunter–gatherers, but during their short lives, these people were (and still are) considered fitter than city dwellers. Contemporary hunter–gatherers are often hailed as the fittest people alive and it seems that the more isolated from modern society they are, the leaner and more muscular they tend to be. This rationale has logic: life is far from easy in the natural world and hunger forces hunter–gatherers off their grass-tuft sofa in search of food. The closer people get to modern creature comforts, the less effort is needed to fill up on food and the softer the sofa becomes, favouring a lazy couch-potato lifestyle.

There are now only a few communities around the world living in this traditional way. The areas left to the hunter–gatherers are usually those where farming is impossible or uneconomical: deserts, dense forests, remote islands and the Arctic. The Pila Nguru people inhabit the Great Victoria Desert in South Australia; they also go by the alias of the Spinifex people, named after grasses prevalent in their locale. Although remaining culturally segregated from the modern world, their biographies have been published and exhibitions of their vibrant artwork have travelled around the globe, reaching London galleries in 2005.[7] The artwork is also available to buy online. While the integration of the indigenous Australians into industrialised living is virtual, obesity remains scarce; once integrated, the tables turn and their susceptibility to unhealthy body weight and other diseases of civilisation is much higher than amongst the descendants of the more recent European arrivals to the country: so much higher that their average lifespan is two decades shorter.

The Sentinelese remain segregated and are yet to cash in on the commercial goldmine of their uniqueness. The situation is unlikely to change. Their habitat, the 72-km square Andaman Island located in the Bay of Bengal, could justifiably be plastered with hazard warnings against making contact: historically the Sentinelese expressed gratitude for gifts of friendship with showers of fire-blazing arrows. Back in the 1970s, when observers tried to make peace, half of the Sentinelese couples were either expecting or had dependent children. From these observations the birth rate is assumed to be close to the maximum and yet the population is estimated to be small, ranging between 50 and 400 individuals, and probably fluctuating within this range as epidemics of disease and famine cycle into times of plenty and back again. There were no signs of obesity or even plumpness among the individuals observed. Since 1990, attempts to study the population at land level have been abandoned after risk assessment showed that the gain of knowledge was not worth the lives lost in its pursuit. Their hostility may be backed by good reasoning: it is not just diseases of civilisation to which formerly isolated populations fall foul on integration into the modern world; many of the other Great Andamanese tribes died out after contact, because of the infectious diseases that hitched a ride on their visitors. Some of those who survived this unintentional cull continue to live in the traditional way, but their culture has been tailored to attract tourism and so, like the Spinifex people, they are no longer true a model of Stone Age life.

The antisocial nature of the Sentinelese protects their culture, but makes it somewhat difficult to learn about their lifestyle and traditions. Still, the limited attempts at contact made during the 20th century were sufficient to suggest that these people are probably the best modern representation of early Stone Age life on the planet today. They adopt a pure hunter–gatherer society with no signs of agricultural practices. Certainly, their culture cannot follow the same evolutionary pathway as in the rest of the world because the island lacks natural resources. Yet these restrictions do not stop the salvage of bounty from the beach, most noticeably when two international container ships ran aground on the island's external coral reefs in the late 1980s.[8] The people hammered metal scraps into new shapes, including weapons.

More detailed observations of the hardships of the natural world have been made in less isolated (and less hostile) communities. The Hadza ethnic group lives in north-central Tanzania around Lake Eyasi in the central Rift Valley and on the neighbouring Serengeti Plateau. Unlike those who live in the Sentinelese habitat, here the population is not limited by watery borders, yet there is little growth in number or in girth. Women produce babies as often as their health and nutrition allows, but about one-fifth die within their first year and nearly half of all children do not make it to the age of 15. Their main foes stem from the extreme heat and the arid desert; frequent thirst needs to be quenched at water pools swarming with tsetse flies and malaria-carrying mosquitoes.[9] These modern hunter–gatherers may be physically fitter than your average city dweller, but they do not have the resources, particularly food and medicines, to enjoy long lives. For Stone Age humans with more limited medical care survival is expected to have been just as difficult. In all likelihood, poor nutrition and poor medical care kept the average Stone Ager too trim to survive for much more than four decades; just like their modern equivalent.

2.3   NATURAL LIVING

Hunter–gathering allows little control over food availability, and natural storage technology does not necessarily allow for healthy provisions all year long. At the beginning of human time, our ancestors hunted for meat and gathered plant foods from their environment. They lived off the fat of the land, but the land was rather lean and the fat was widely dispersed. For these reasons, travelling was a significant part of hunter–gathering and, as Stone Age travelling was on foot, early humans knew the true meaning of endurance exercise. Survival made them push themselves to obtain a truly athletic physique, and in the process burn fat and lower cholesterol levels. Still, travelling was only part of the energy expenditure involved in making a meal.

Apart from our intelligence, we humans do not have many survival attributes. Our physical design does not help with hunting: we do not have sharp teeth, claws or speed like other hunting animals. Consequently, trapping and hunting require planning, skill, dexterity, speed, concentration, a lot of time and, last, but by far not least, tools. Similarly, gathering plant foods is not as easy as it is in a farmed environment. Today and then, strawberry picking may be a family affair, but whereas we are now able to select from the modern large, juicy fruits planted together in nice neat rows, back then it involved some serious walking to find the widely dispersed fruits, coupled to sharp-eyed spotting because wild strawberries are only the size of a pea. Without advanced tools the collection of all foods was slow, hard work. Gathering grains involved painstakingly picking each seed from the grass stalks or scooping the seeds from the ground. Vegetables needed digging from the dirt or cutting above it with a flint knife. Although technology evolved early,[10] a hunter–gathering existence required skill, knowledge, endurance, exercise, a lot of patience and even more energy. The very process of meal-making mounted many blocks on the energy balance to off-set the calories in the food before they even entered the body.

The labour involved in meal-making did not end with the collection of the food. Freshly found foods were often inedible and needed preparation before they could become part of supper. Animals needed butchering, grains needed grinding, nuts needed shelling, beans were often dried to boost their storage potential or cooked to aid digestion, and vegetables were very tough, requiring plenty of chewing or plenty of cooking before they were suitable for swallowing. Chewing tough food wore our ancestors' teeth down to the gum by the time they were in their thirties; toothlessness was not caused by rot associated with acids and sugars as it is today, rather it was caused by wear and tear. Without a local dentist to repair the damage or offer a set of dentures, cooking and food processing became an important health investment.

The benefits of food processing extended beyond dental health, because many wild foods carry poisons; acorns are a prime, but not exclusive, example. Nearly all natural fruits and vegetables are laced with one special chemical or another made by the plants for their own purposes. These phytochemicals have different effects, depending on the species and size of the animal they enter. Some are toxic to us; others just kill bugs. In the latter case, at the right dose some plant products can benefit human health. For example, brassicas are bitter because they produce effective antibiotics designed to protect the plant from bugs, and these same chemicals kill the bacteria that cause the stomach ulcers that can lead to cancer. These bacteria, given the scientific name *Helicobacter pylori*, are becoming resistant to commonly used synthetic antibiotics, yet eating Brussels sprouts is still effective. In contrast, other plant toxins are deadly poisons; some are famous tools used in Victorian murders. Cyanide is produced naturally by cherry trees in their twigs, leaves and seeds to make sure that diners focus only on the specially made fruit and not the tree itself. The same is true for apples, almonds, apricots and peach trees. The amount of poison in an apple pip will not kill a human, but when a smaller animal makes it their meal, the impact is more significant. On the other hand, just a few pips or seeds from other plants can be fatal for humans. For example, foxgloves are notorious for seeds that cause heart-spasm. Other plants seem kinder in that they make fruits specifically for animal consumption in a bid to distribute their seeds, which are also specially adapted to pass straight through the gut. Tomatoes are an example. However, to help with the passage, plants lace the meal with natural laxatives. Speedy passage reduces the amount of nutrition absorbed from a meal.

Therefore, plant foods have a significant effect on human health, causing on one end of the scale diarrhoea and on the other instant death. Is it any wonder that humans have inbuilt food preferences that favour the consumption of meat? These preferences not only protected against poisoning, but also maximised energy consumption. Given the hard slog of survival and low food availability, the energy balance in the Stone Age was tipped in favour of the lean physique envied by modern city dwellers. However, many of these people had too little fat to balance their lifestyle, which is part of the reason their average life expectancy was what we now call middle age.

2.4 TAKING CONTROL

The practice of farming – the second stage of nutritional history – not only set out to stabilise the quantity of food available, it set out to stabilise the quality too. For the hunter–gatherers, Mother Nature preselects the range of foods on offer and then, guided by genetically programmed taste preferences, the people select the foods that best match their needs. This process is effective because, although some food preferences are developed personally by culture and experience, most are a fundamental part of human nature.

Galanin is one of several signallers involved in population-wide food selection. This signaller was first discovered in extracts of pig gut in 1978 by Professor Viktor Mutt and colleagues at the Karolinska Institute, Sweden.[11] Before this contribution to science, signalling in the body was mainly thought to involve specific structures called glands: the hypothalamus, thyroid and pituitary are examples. However, Mutt's work revealed the roles of other parts of the body in healthy signalling – body parts that had already been assigned other functions. He identified many peptide signallers released from the gut that regulate digestion, suppress hunger or stimulate the same. Other researchers found more: so many that scientists decided that a new sub-discipline needed to be defined and "gut endocrinology" was born. The birth of the new sub-discipline gave the gut gland status; it also opened the flood gates for the understanding of fat chemistry.

Mutt has been described as "the scientist who has single-handedly contributed the most to the development of gut endocrinology" by the International Regulatory Peptide Society.[12] He demonstrated that, for the endocrine signallers, size is not everything. Very powerful effects can be mediated by relatively small molecules, such as galanin. Galanin is made up of the same building blocks as proteins – amino acids – but it is much too small to be a protein. Proteins are chains of more than 50 amino acids, strung one after another. The chains coil and bend to form complex yet highly reproducible structures, which are important for protein function. Structure is important for galanin's function too, but with only 30 amino acids in its chain, it misses the protein categorisation cut-off. Short chains of amino acids, such as galanin, are known as peptides and are often prefixed with their function. Galanin is a neuropeptide, used by nerves cell (neurons) to communicate with each other. Specifically, it dulls nerve signals at a molecular level, and its general nature means that it touches many regulatory pathways. As such, malfunctions have been linked to many disease states. For example, changes in its functions have been linked to Alzheimer's disease, but it is unclear whether its role is in aggravating the symptoms or compensating for the characteristic damage.[13] In support of the second hypothesis, galanin has been implicated in recovery of nerve cells after damage caused by trauma.[14] Its role in neurodegeneration (or perhaps protection) may be ambiguous, but its role in appetite regulation is clearly defined: galanin stimulates the appetite. Unlike some other appetite-stimulators, it does not mediate its effect in response to shortages; rather it responds to the nutrients the body receives. It boosts appetite in response to dietary fat and alcohol, and the appetite boost is not for any old foods, but for more of the same.[15] In this way, galanin is part of our genetically programmed greed and taste preferences: it makes us greedy for high-energy foods.

Favouring the flavours of high-energy foods and linking their consumption to appetite stimulation – and also pleasure – may have helped our ancestors to survive. These foods are rare on Mother Nature's menu, but they help offset the endurance exercise that was an essential part of life. They provide vital nutrients and a much-needed energy boost for natural living. Galanin's effect on appetite is quite specifically tailored to such a role: rather than craving high energy foods all the time, the desire is only initiated after these morsels have entered the system; the demand is initiated by the supply. In this way, our ancestors were largely content with what they could find, but when energy-dense foods were available, they ate their fill, perhaps even beyond their fill.

In the Stone Age, the interplay between desire, natural living and laziness provided the most appropriate diet from the available food. It was not ideal, but the best given what was on offer. It was successful with respect to both quality and quantity. The quantity factor is easy to gauge from the degree of general hunger and it seems that prehistoric humanity was resourceful enough to deal with that problem by whatever means available, including harvesting acorns. The quality factor is less easily measured without chemical know-how, but, somewhat surprisingly, our ancestors did a super job of obtaining nutritional balance, at least when food availability forced them to eat plant foods. Modern evidence for their success is found in the nutritional balance provided by many traditional food combinations. For example, both beans and grains lack one or another essential amino acid, but together they provide the complete range; beans and grains were ground together to make bread in the Stone Age and this practice prevented the nutritional disease pellagra. The power of these instinctive food choices is also demonstrated by the fact the human race is still here: despite having few physical survival attributes and being challenged by cold, hunger and limited food supplies (widely distributed between similar-looking but quite poisonous alternatives) our ancestors managed to reproduce. They used their innate programming and intelligence to select the foods that matched their active existence. With such powerful instincts guiding our food choices is it any wonder that farming favourite foods became more popular than relying on Nature's generosity?

Farming offered an alternative to picking meals from Mother Nature's preselected food range. It allowed humanity to take the role of the selector and increase the supply of the foods favoured; it allowed us a new freedom to indulge food whims, but this practice was still under the overall power of Nature. The popularity of farming is likely to be rooted not only in the quantity of foods offered, satisfying hunger, but also in the relative quantities of different foods, satisfying preferences. Just as today shopping baskets can be filled with fruit and vegetables or energy-dense junk foods, the early farmers could choose whether to make use of the land to produce more meat, more cereals, or more vegetables;

innate food preferences, including the work of galanin, favoured the first two food groups.

Once farming was introduced it allowed the population density to exceed the land's natural capacity for providing nutrition. This growth may have heightened the risk of starvation when harvests failed – the bigger the community the harder it falls – but such digressions did not detract from the goal of farming foods and, even when forced to dip back into Mother Nature's pantry, the people returned their focus to farming as soon as they could. The lure was strong. Many historians think that early farming, with its lack of tools and technology, took more effort than hunter–gathering; its success suggests it had an attraction that overcame laziness. Perhaps it provided for the more pressing need of hunger, but it also indulged genetically programmed food preferences. Certainly, when farming began our ancestors lacked protein and fat in their diet and it offered a way of meeting those needs.

One hundred times as many people can be supported from the same area of farmed land compared with foraging. As the human population reached its capacity for the hunter–gathering lifestyle, some bright spark realised that the food supply could be stabilised if the land was only used to grow the plants and feed the animals that the people wanted. Starving people could be fed by farming and, as this practice spread, population growth become more and more rapid throughout the second stage of human nutritional history. With farmed foods, people could live in bigger communities, perhaps benefiting from the protection that these larger numbers offered. Bigger communities also allowed more technology to be developed. Walled cities became possible, as did ploughs. Animals could be roped into the labour of farming, which was simply not possible to the same degree in hunter–gathering. And then machines replaced animals. With each step more people were freed from food production to find other careers. Art, science, technology and medicine became livelihoods rather than hobbies. The future looked bright.

2.5 THE BEGINNING OF THE END

As soon as people started to move out of the natural world, their dietary balance changed and the second stage of nutritional history began. In the natural world, food availability was balanced with taste preferences and laziness to give a diet that was about one-third meat and two-thirds vegetables. Nutritional needs have now changed such that this diet is no longer healthy, but many people continue to eat in this way. A daily Stone Age ideal meal plan may have been something along the lines of: for breakfast, equally sized servings of clams, cabbage and corn; lunch, half a chicken and the same serving of each of chicory and chick peas; dinner, a similar sized steak, salad and serving of sweet potatoes. It is essentially the British main meal of meat and two vegetables three times a day. Such a high meat content would not be healthy in modern cities, but for the very active hunter–gatherers, whose lifespan was limited by insufficient nutrition and poor medical care, this diet was about the best they could select from the foods on offer. It provided essential fat and protein without causing obesity. Farming changed the balance.

Farming allowed innate taste preferences to be increasingly indulged and it allowed the human diet to shift to include even more meat and grains at the expense of vegetables. Farming eventually allowed meat and two carbohydrates to become the main meal: a burger, a bun and some fries.

As farming began, there was no source of scientific data to guide people in their food choices and there were no officials to provide the public with unbiased information on health-optimising diets. Our ancestors relied on instinctive food preferences. We all have our favourite foods, but some are common to most people – the sweet and fatty flavours. Guided by these taste preferences, farmers focused on making meat more and more obtainable; they reduced the energy, time and effort necessary to acquire our favourite foods and, in so doing, the effort gap between different foods diminished. Still, our food favourites remained the same. And so as the second stage of nutritional history advanced, our ancestors increasingly ate energy-dense foods. The change in the relative availability of meat and vegetables reduced the power of our genetically programmed instincts to promote a balanced diet. Our ancestors continued to follow the instincts that had, thus far, served them well without realising the health decline that their nutrition was causing.

At the beginning of the second stage of nutritional history – the farming revolution – there were no distinguishable detrimental health effects as a consequence of following our instincts; they seemed to continue to serve well. The Stone Age population size was likely to be limited by energy, and so was life expectancy. Planting energy-dense cereals provided a way of harvesting more food-energy from the same amount of land. The selection process did not stop at choosing which type of food could be produced; longer-term projects focused on tailoring food molecularly to meet our whims. Our ancestors selected plants and animals best suited for their needs and bred them to produce increasingly specialised varieties. As early farmers selected the most useful plants to propagate the following year, a new chapter began in the history of the world, where people took the place of Mother Nature in selecting the species that were most fit to survive in *their* world. The characteristics that were useful for humanity were often not the ones that were useful for natural living, and many domesticated species now need human intervention to survive. For example, take one of the first cereals, as cultivated grasses are known, called einkorn – the type of cereal that Ötzi's folk cultivated. Einkorn originates from the Near East. The wild plants systematically shed their seed on the slightest wind or brush of a passing animal, from the top of the ear to the bottom over a period of weeks. The slow and sensitive seed release maximises the chance that these seeds will produce mature plants. Unusual weather patterns may destroy all the seedlings at any particular time, but if the seeds are shed slowly there is more chance that some seedlings will be strong enough and others not yet released when any bad weather strikes, increasing the chance that some plants will grow to maturity to provide the next generation. Slow shedding also reduces the direct competition between seeds from the same plant. However, after many generations of human intervention, domesticated einkorn has changed to the benefit of humanity: all the grains ripen at about the same time and the seeds do not come off until the harvest is threshed, maximising the amount of grain collected. This altered shedding pattern was the result of a gradual shift in the varieties of genes within the domesticated einkorn population. With each generation, breeders select for or against different characteristics, which in turn selects for or against the inheritance of different varieties of the genes that determine the selected feature, changing the designs of wild plants and animals. This selection has produced all the domesticated species.

Interfering with the patterns of growth alters the gene pool – the varieties of each gene represented in the entire population – and, therefore, with farming came a crude form of genetic engineering that gradually changed the texture and taste of foods obtained straight from the field. The impact was so sensational that, in addition to the spread of the farming lifestyle, many of the species that had been grown in the Fertile Crescent were exported. Certainly, this combined package arrived in Ötzi's homeland, such that by the time of his death non-indigenous species were being cultivated. Imports aside, the dietary change was a slow drift, taking decades to have much effect, but over the centuries since farming began the cumulative effect has been significant. Meat gradually changed from strongly flavoured game to blander, more succulent farmed cutlets and yet, while traditional farming methods were employed, nutritionally it was more or less the same. Vegetables drifted away from their tough, strongly flavoured origins towards a more palatable mild softness. Still, even with these improvements in palatability caused by selective breeding, the popularity of vegetables fell as meat and cereals became more readily available. Dietary shift was in part attributable to the work of peptide signallers, such as galanin, operating against low resistance, due to a lack of scientific understanding.

Whether it was the farming technology alone that migrated across the world, in the form of seeds and know-how bought from traders, or whether people emigrated from the Near East, taking the technology with them and gradually pushing the other cultures out of existence, is not clear. However, it is likely that the final force in favour of farming was the devastating consequences of this practice on the natural environment. Farming is thought to have put pressures on the resources that were necessary for hunter–gathering, squeezing that lifestyle out of existence.

2.6 ADULTERATION

Farming did not start immediately to develop into is the form it takes today; the process was slow, taking generations to breed plants and animals selectively to make them more suitable for our wants (note 'wants' rather than 'needs'). The process of selection and improvement was slow and methodical, designing animals and plants to produce more and more food, with less and less effort from people; which appealed to our in-built greed and laziness. From when farming began, right up to the Middle Ages, families in

Britain were virtually self-sufficient. An agricultural society was made up of many small villages that had yet to experience the local convenience shop, or really any sort of trade in food. The members of each household cooked their own food, baked their own bread, butchered their own animals (and probably fattened them too) and brewed their own beer. However, as time went on gradual technological improvements in farming and food production worked towards the goal of providing more food with less input from humans: the plough and the mill are examples. Increased technological investment spurred the division of labour. Perhaps almost every family had their own plough, but some would share; and those that did could afford a better model. With increased use of the same tools came a higher level of skill, and trades developed; rather than hiring the tools, people traded their expertise for that of others. Step by step food production became one of the many jobs necessary for survival that was more efficient when outsourced.

When families were self-sufficient, they needed enough land to provide their food: a field or two to grow some crops and fatten some beasts. As time progressed and the labour was divided, bread could be bought from the baker, meat from the butcher and fish from the fishmonger. And, as long as land was not needed for their trade, people could live closer together. Cities offered a whole range of benefits, including reduced transport and overhead costs for producers, and therefore cheaper products for the consumer, protection from attack (many early cities were walled), and a more diverse culture. As small cities appeared, art, technology and science advanced. Living close to people working on similar problems allowed discussions that led to like-minded individuals coming up with new theories and solutions. Many breakthroughs freed more time, generating a cycle of one breakthrough in science, technology or medicine that created more time to make more breakthroughs, improving the quality of life until today's living standards were reached and more free time than ever was created. The future still looked bright. But the city dwellers' health suffered. Their average life expectancy was little more than the four decades the hard-pushed hunter–gatherers had enjoyed, which was surprisingly short given the improved standards of medical care. Infant mortality was shockingly high and the most common disease was chronic gastritis, suggesting that food was part of the problem.

As farming was established, food selection was governed less by need and more by choice, and as the choice was driven by innate programming, people ate more energy. With crop selection and selective breeding focusing on calories, the energy density of the available foods began to rise. Wealthy people could better follow their genetically programmed instincts: meat became the main focus of food, padded out with bread and, in the last 400 years or so, potatoes. Poverty restricted choice, forcing modest vegetable consumption and inadvertently improving health. The nutritional deficiencies among both the rich and poor that resulted from such a diet were hidden by a more sinister problem that had arisen from the combination of free time and human greed: food adulteration.

When people were self-sufficient they were absolutely in control of their food and they knew exactly what went into every stew and loaf. Because they cooked for themselves, they used the best ingredients available, which was not always the case if food was bought from outside the household. When profit was introduced to food production, corners were cut and health suffered. Some of the less problematic examples of food adulteration included watering down milk or mixing coffee with cheaper chicory.[16] However, there were more stomach churning examples to come. A substance known as alum was added to bread, to make it look whiter, as though it was made from the more desirable highly refined flour. Alum is a white mineral salt that contains aluminium and potassium sulfates, which was used at the time to whiten paper. In bread, this chemical inhibits digestion, lowering the nutritional value of the meal to such an extent that this type of adulteration has been pinpointed as a cause of the rickets epidemics of the past.[17] Although very little aluminium is absorbed from food, once inside the body it can accumulate in the brain and it has been linked to neurological disorders including Alzheimer's disease,[18] but our ancestors did not live long enough to worry about degeneration, mainly because of their diet. Various cheap additives were put in beer in place of malt and hops and were used to disguise the taste of watered-down ale. One of the additives was a poison made from berries of the Indian *Anamirta cocculus* plant,[19] a narcotic and a stimulant. With these properties, adding the berries to diluted beer would still make customers intoxicated, perhaps even more so than had they drunk unadulterated beer! But the health problems associated with this kind of poisoning were more severe than those caused by consumption of the expected amount of alcohol. These berries were used in an ointment to kill lice and were traditionally thrown in water to stupefy fish. If poisons in beer and bread do not put you off your meal, there were reports of cream thickened with animals' brains;[20] poisonous copper and lead carbonates added to some foods, which can cause kidney damage and mental retardation; sweets coloured with toxic mineral dyes; lead added to cayenne pepper; and rancid meat spiced to disguise the taste. As the British public flocked to the cities to be part of the new workforce during the industrial revolutions, they increasingly relied on food producers, out of both necessity and convenience. Ready baked breads, pies and filleted meat cuts were the convenience foods of the time and the poisonous load introduced by food adulteration took its toll.

With time, science became sophisticated enough to identify some of the food producers' secret ingredients. By 1820 things were shown to be so bad that a German-born chemist, Friedrich Accum, published a treatise on the culinary poisons which declared *"There is death in the pot"*.[21] Accum showed that adulteration of food was enormously common. In fact, one of the reasons why infant mortality was so high in the cities was because of the poisons in food.[19] The damage caused by each dose added up and some poisons interacted with each other. The addition of lead, copper, mercury, arsenic and other deadly ingredients by our ancestors to each other's food was the reason why chronic gastritis was the most common disease of the urban population in the 19th century.

Despite Accum pointing out the problems with food production and how they were affecting the health of the population, those who could implement change remained blissfully unaware of the situation, until 30 years later when a radical doctor called Thomas Wakeley demanded an investigation into food adulteration in the UK, bringing the situation to the forefront of public attention and forcing the Government to take action. The action came in the form of an enquiry. Dr Arthur Hassall, a physician and lecturer at the Royal Free Hospital in London, was appointed to conduct it. His report showed that if there was money to be made from food adulteration then food was being adulterated.[22] It was virtually impossible to buy any basic foods that had not been tampered with.

More importantly Hassall, for the first time, published the names of the shopkeepers and producers that provided the tampered food to the public. This report allowed people to see how such practices were affecting them: for the first time people could tell whether their local baker was adding alum or iron to their bread. As the lists grew and more and more people were affected, food adulteration became a parliamentary issue and the Government began to take more interest. Questions were asked in parliament as to why there was no legislation regarding the food safety. The final report concluded that food adulteration was fraud, and a fraud that was affecting people's health – the most valuable thing they possess. Four years later the Food and Drugs Act became law. Legislation was passed in the UK in 1860 and 1872 that encouraged the appointment of local food analysts and imposed penalties for tampering with food. The legislation was part of the transition into a new era of human nutritional history.

The third stage of nutritional history is distinguishable from the first and second by two sets of scientific advances that converged: those that allowed legislation to protect against the addition of poison to food and those that allowed the molecular details of nutrition to be mapped. As science and technology improved, the nutritional needs of the population could be defined and the chemical composition of food could be probed to determine whether it matched these needs. With an accelerating rate of scientific development, it was only a matter of time before the advances reached a level where food products could actually be designed to meet these needs. In the last half of the 20th century the effects were sweeping: food technology advanced more rapidly than the developments in nutritional science. The rapid change was, like farming, directed by our genetically programmed taste preferences rather than our intelligence. It represented, in keeping with the first and second stages of nutritional history, a progressive loosening of restraints on food production, but this third stage had a chemical weapon to break down any barriers that impeded the satisfaction of our innate preferences. By the end of the era, home-cooked foods no longer tasted better than those that could be bought, and they were no longer deemed healthier: manufactured products came with a defined nutritional composition and some began to be supplemented, apparently improving their value. Nutrition seems to have been distilled to 20 or so chemicals listed on the side of packets. But why read this information when a multivitamin supplement provides all that is needed? Or does it? The rate of change in the third stage was more rapid than the others, but the results were similar: widespread nutritional problems caused by an unbalanced diet.

REFERENCES

1. K. Connolly, in *Electronic Mail & Guardian*, 21 August 1997.
2. M. Uhle, *U. Calif. Publ. Am. Archaeol. Ethn.*, 1907, **7**, 1–106.
3. T. L. Stedman, *Stedman's Medical Dictionary*, Lippincott, Williams & Wilkins, Philidelphia, 28th edn, 2005.
4. K. Choy, O.-R. Jeon, B. T. Fuller and M. P. Richards, *Am. J. Phys. Anthropol.*, 2010, **142**, 74–84.
5. F. Ivanhoe and P. W. Chu, *Int. J. Osteoarchaeol.*, 1996, **6**, 346–381.
6. F. Ivanhoe, *Int. J. Osteoarchaeol.*, 1995, **5**, 213–253.
7. S. Cane, *Pila Nguru – The Spinifex People*, Fremantle Arts Centre Press, 2002.
8. S. A. Awaradi, *Andaman and Nicobar Adminstration*, Port Blair, 1990.
9. http://ngm.nationalgeographic.com/print/2009/12/hadza/finkel-text (accessed 3 July 2011).
10. M. D. Leakey and D. A. Roe, *Excavations in Beds III, IV and the Masek Beds, 1968–1971*, Cambridge University Press, Cambridge, 1994.
11. D. Wynick, S. W. Thompson and S. B. McMahon, *Curr. Opin. Pharmacol.*, 2001, **1**, 73–77.
12. http://www.regpep-society.com/Viktor-Mutt.html (accessed 3 July 2011).
13. S. E. Counts, S. E. Perez and E. J. Mufson, *Cell. Mol. Life Sci.*, 2008, **65**, 1842–1853.
14. R. P. Hulse, D. Wynick and L. F. Donaldson, *Mol. Pain*, 2011, **7**, 26.
15. S. F. Leibowitz, *Neuropeptides*, 2005, **39**, 327–332.
16. H. Mayhew, *London Labour and the London Poor*, Dover Publications, New York, 1968.
17. J. Snow, *Int. J. Epidemiol.*, 2003, **32**, 336–337.
18. V. B. Gupta, S. Anitha, M. L. Hegde, L. Zecca, R. M. Garruto, R. Ravid, S. K. Shankar, R. Stein, P. Shanmugavelu and K. S. J. Rao, *Cell. Mol. Life Sci.*, 2005, **62**, 143–158.

19. F. Accum, *A Treatise on Adulterations of Food and culinary Poisons*, Longman, London, 1820.
20. A. G. Payne, *Cassell's Vegetarian Cookery – A Manual of Cheap and Wholesome Diet*, Indypublish.com, USA, 2005.
21. C. A. Browne, *J. Chem. Educ.*, 1925, **2**, 827–851.
22. A. H. Hassall, *Food and its Adulterations; comprising the reports of the analytical sanitary commission of 'The Lancet' for the years 1851 to 1854*, Longman, London, 1855.

CHAPTER 3

Left to Our Own Devices

Whereas in Ötzi's era nutrition kept the body mass index (BMI) at about $18\,\mathrm{kg\,m^{-2}}$, by the time the Hanoverians were sitting on the throne in England, the average BMI of the population was nearer to $22\,\mathrm{kg\,m^{-2}}$. The survival story had been softened by improved medical care, but poor nutrition – deficiencies and excess – kept longevity in the same domain as it was in the Stone Age. Throughout the second stage of nutritional history humanity had more control over the range of foods available and the choices they made revealed a truth about human nature: it demonstrated that when Mother Nature loosens her grip on the food supply we follow genetically programmed taste preferences and opt for salty, sweet, fatty and protein-rich products. But through much of human history we did not overeat. We are not destined towards obesity. The feelings generated by eating our favourite foods promoted them to medicinal status – food and mood are closely intertwined. Hippocrates, the founder of modern medicine, is famously quoted as saying words to the effect of let food be your medicine and medicine be your food. His other writings suggest that when his patients were sick, diet was considered as a primary suspect, and when healing was required, nutrition was manipulated. His dietary recommendation for good health was to eat modest amounts of fish and vegetables. However, with time Hippocrates' wisdom was lost – or perhaps ignored. People began to lean on their own understanding of

Fat Chemistry: The Science behind Obesity
Claire S. Allardyce
© Claire S. Allardyce 2012
Published by the Royal Society of Chemistry, www.rsc.org

food and health, which was largely based on the same genetically programmed preferences that had spurred the development of farming and, as a result, there were epidemics of nutritional problems, yet obesity was not one of them.

By the 16th century, the second stage of nutritional history was in full swing. Most people subsisted on a diet of meat and cereals washed down with mead; fruits and vegetables were avoided and in 1569 their sale was banned on the streets of London: the people had mistakenly linked sickness that resulted from eating unclean fruits with the plague. Vegetables were food for the poor, with the rich opting for mainly meat, bread and occasionally a carrot. But in the midst of this scientifically unfounded anti-green campaign, Andrew Boorde published a book on dietary health called *Compendyous Regyment or Dyetary of Health* that suggested, amongst other things, that eating meat helped people to become strong, with veal being easily digested, 'brawn' (boar's meat) being a fine winter food, and bacon was an appropriate meat for carters and ploughmen. He seemed to have identified protein as responsible for body building and correlated the energy needs of manual labour with the fat in the food, yet his correlations lacked scientific substantiation. Boorde was before his time in other ways too. He recognised the importance of vegetables, suggesting that his readers ate from a modest selection of plant foods: onions, leeks, garlic, and apples. This list of four marked the beginning of an important turning point, going against genetically programmed taste preferences to improve health. Innate programming no longer served us well. Nutritional deficiencies were rife; as were signs of certain excesses. In addition to the transmissible diseases and consequences of adulteration, gout, oedema and unwanted stoutness plagued the population. Life expectancy was about 47 years; less in cities where poor sanitation also took its toll. Food began to be suspected as the root of the problem, but the horrific details could not be characterised without advances in chemistry that were not due for another 200 years; in the meantime the focus on nutrition shifted from quantity to quality, favouring foods renowned for improving health, those that were energy dense.

Leading up to this period, the Old World had been strongly influenced by Renaissance Europe. In England, this was the era of King Henry VIII and Queen Elizabeth I pulling Britain out of the Middle Ages and into the modern world, Shakespeare's plays,

Hilliard's paintings, and a burst of social activity, development and trade. The social activity brought with it fabulously decorated apartments, magnificent costumes and elaborate ceremonies in the courts. Amongst all this extravagance, food was considered of great importance – a symbol of wealth. But obesity was for the poor: although Henry VIII appears to have been on the pudgy side, his wives and daughters were bound, by corsets, to be thin.

Despite the fatty diet, Britains living under the reign of Elizabeth I were trimmer than those under Elizabeth II, and not just because of corsets. Food production increased throughout the second stage of nutritional history and many people had the resources to advance up a BMI zone or two, but they chose to eat less than the ideal and stay fashionable. And hard work helped their cause. Hard work remained an integral part of living for the majority of the population right up to the modern day. Before the sequential information technology revolutions that began in the 1950s, most people were involved in manual work and manual work is physically demanding: it involves moving, lifting and sweat. For many, the body received a full workout from employment alone – even if the employment was not salaried. Housewives (as it was invariably wives that stayed at home) worked hard all day cleaning, washing and sweeping without the luxury of the products and machines that make modern living much easier. On top of that, they had to raise a handful of children and tend a vegetable plot in the garden. Daily activities mounted block after block on the energy balance, so when people found free time, they rested. They rested without significant weight gain because the effort of living itself seems to have offset their meals and the energy consumption was balanced.

Superficially, hard work seems to have allowed our ancestors the privilege of being able to eat what they wanted, at least when it was available. Following genetically programmed food choices, right up until the beginning of the 20th century our ancestors opted for sausages, bacon, fried bread, suet dumplings, spotted dick and rice pudding. Yet it was not just work that kept our ancestors trim: for the majority, it was the size of the portions they ate that prevented weight gain before the onset of middle-age spread; indeed, most did not even manage to eat the current recommended daily calorie allowance. In Britain, throughout the 1950s, the shortfall was estimated to be about 6.5%. It is hardly surprising then that these people were slimmer than their heirs as they challenged obesity from both sides of the energy balance: they took more exercise and ate less than what is currently recommended. Still, they were heavier than one would anticipate given this balance. Middle-age spread was common, and the population was not healthy; there were examples of debilitating nutritional deficiencies and excesses. In the latter camp, heart disease was a major blight. Chemistry has now revealed how fat and cholesterol not only damage the heart, but cause a whole range of health issues including priming the next generation for obesity. Despite the fat they ate, from the 16th century to the fabulous fifties fashionable people remained slim.

3.1 A NEW FOCUS ON HEALTH

The links between fatty foods and health problems were proven by the scientific developments that mark the third stage of human nutritional history – the period when synthetic chemistry got a grip on food technology. A series of improvements in science precipitated a new focus on food: diet was brought to the forefront of people's minds because medical technology and sanitation had improved so much that the effects of malnutrition became more noticeable and less acceptable. Some simple and low-cost interventions had major impacts on human health and, as scientists made firm connections between nutrition and disease, health began to improve dramatically.

Just as there is no single date when humans turned from hunter–gathering to farming (indeed, there are still hunter–gatherers today), the second transition was similarly vague. However, whereas the first transition took millennia, the second transition happened over a few centuries. If a single starting point could be defined it would be on a ship in the middle of the 18th century.

The first scientist to receive public acclaim for applying modern scientific methods to the study of nutrition was James Lind. In 1747, Lind showed that including citrus fruit in the diet of sailors prevented the nutritional disease scurvy. His conclusion that such health problems are better prevented than cured applies as much today as it ever did, both to examples that are absolutely dependent on diet, such as scurvy, and those with looser links, such as colorectal cancer. This theory also applies to obesity. Obesity is better prevented than cured – but better cured than left to fester.

Lind's work was one of the first controlled clinical studies. He was looking for a link between the diet and lifestyle of sailors and the onset of scurvy. In order to identify the cause of this disease, he had to consider the effects of many factors: the cramped living conditions, the poor sanitation, the stifling heat, the lack of fresh water and a diet of salted meat, biscuits and rum, plus the effect of any combination of these things. With so many variables to consider, how do you know where to begin? Lind started out by selecting 12 men from the crew of the ship *HMS Salisbury*, on which he was serving as a surgeon. Then, he divided the men into six pairs. Each pair was given a different dietary supplement, for example cider, seawater or portions of citrus fruit. By introducing a simple change into the lifestyle of one group of sailors, with another group carrying on as they were, the effect of the change on the men's health could be gauged. The first group of sailors is known as the "test" group, as Lind would test his theories on these people. The second group is known as the "control". At the outset of the experiment, the characteristics of the men were indistinguishable: they sailed the same ship, ate from the same stock, had similar levels of health and lived similar lives, with the only noticeable difference being Lind's doing. As he approached the cure of the disease, the health of the men in each group became notably different: just one group was healthy.

Such split-group studies are the foundation of many types of experiment, even today. The control group is critically important because it shows what would happen without intervention. Scientific experiments are not considered valid without appropriate controls. In the case of the sailors, they usually developed scurvy. But just in case this voyage was different from the many fateful trips before – and just in case Lind's own remedy actually progressed rather than suppressed the symptoms – the control group acted as the reference point. Back on *HMS Salisbury*, most of the men developed scurvy – in both the control and the test groups – except those given citrus fruit supplements.

Lind's work was so significant that it precipitated the first dietary guidelines put in print: the 1835 Britain of the Merchant Seamen's Act enforced sailors' rations of lemon or lime juice.[1] Lind defined a link that led to a cure, but he was laid to rest long before it was possible to define the chemistry of the mechanism of protection. The active ingredient that prevents scurvy is now known to be vitamin C, and this food component can be produced synthetically so efficiently and at such low cost that there is no reason why anyone in the world should suffer from this type of malnutrition today – but they do.

At the time of Lind's work, vitamin C deficiencies affected many people on terra firma, as well as on long sea voyages, mainly because of the general distrust of fruits and vegetables. Some of these deficiencies could have developed into full-blown scurvy; some may have been fatal. But scurvy manifested itself in epidemic proportions on ships because the sailors could not follow their genetically programmed instincts that protect against malnutrition: they could not satisfy their food cravings just as they could not escape the diet of hard tack and salted pork. The food rations met their energy requirements – at least enough to allow them to work – but did not consider their complete nutritional needs.

3.2 THIRD TIME LUCKY

James Lind began a revolution in scientific studies, which, coupled to technological advances, transformed our understanding of nutrition beyond empirical associations to, eventually, molecular facts. The whole period took almost two and a half centuries. This timeframe may seem long when our lifespan is just three score years and ten, but compared with those over which the first two stages of human nutritional history began, progressed and reigned, it is surprisingly short – short enough for mistakes that were made in the early days to remain. The change happened so rapidly that a misunderstanding could completely derail progress.

Take, for example, the story of the discovery of thiamine. More than a century on from Lind's work and this time in Japan, Dr K. Takaki, then the Director General of the Japanese Naval Medical Sciences, showed that improving the diet of sailors to include vegetables eliminates beriberi, a disease with symptoms such as severe lethargy and systemic organ problems, including inflammation of the nerves (neuritis). However, unravelling the molecular cause of the disease involved the work of several scientists and some chickens. A Dutch scientist, Christiaan Eijkman, picked up where Takaki left off. He studied beriberi in what is now Indonesia towards the end of the 19th century and made a major advance when he noticed that, in addition to his patients, the fowl on farms outside the hospital fell sick with the same symptoms. Closer inspection revealed that chickens fed on white rice developed the disease, but those fed on whole rice were fine; the processing was having an effect on the birds' health. To explain his observations, Eijkman contrived an elaborate hypothesis in which the white rice contained a bug that made the chickens sick and the outer coating contained an antidote such that birds fed on whole rice stayed healthy. A few years later, his colleague was able to extract what was thought to be the antidote, but its identity remained elusive until 1926 when a new team in the same laboratory in Indonesia collected data that gave them an insight into the molecular details of the cure. Unfortunately, they made a mistake in identifying some atoms and their positions, which caused much confusion until 10 years later when Robert R. Williams, an American chemist, corrected the error and the active ingredient was identified as vitamin B_1, or thiamine. He went on to make this nutrient synthetically and his work was honoured with the Elliott Cresson Medal in 1940 and the Perkin Medal in 1947.

The mistakes Eijkman and collegues made could have derailed the understanding of the contribution of thiamine to health, but a new light was beginning to shine over in England: the biochemist Frederick Hopkins categorically proved the importance of what he called "accessory food factors". In restricting the diet of dogs to just energy and protein, he showed scientifically that they became sick.[2] So groundbreaking were these contributions to the understanding of human health that Eijkman and Hopkins were jointly awarded the 1929 Nobel Prize in the domains of physiology or medicine: Eijkman "for his discovery of the antineuritic vitamin" and Hopkins "for his discovery of the growth-stimulating vitamins". The antineuritic vitamin was vitamin B_1 (thiamine). Just as in the case of the growth-stimulating vitamins, its discovery was via the devastating health consequences of deficiencies.

In the body, many proteins are categorised as enzymes. These proteins are catalysts: they accelerate the rate of chemical reactions without being consumed. Chemical reactions need to be accelerated in the body to support life. Thiamine is processed by our bodies to form thiamine pyrophosphate, which helps specific enzyme machines do their jobs; it is a co-enzyme. It does not catalyse the chemical reactions, but helps the enzymes to act as catalysts and is often modified in the process, requiring regeneration. The enzymes it works alongside include some involved in the breakdown of glucose. Therefore, in order to process the starch in rice and other carbohydrate-rich foods effectively, the body needs the thiamine in the grains' outer coatings. Thiamine also helps one of the enzymes that is involved in DNA synthesis and is found in our brains and nerves, although quite what it does there is not clear.

Rice is processed because the brown outer coatings make the grains prone to spoiling, and when they are removed the extended shelf life reduces the cost of the food. In addition to thiamine, the outer coating contains several other vital nutrients, along with trace amounts of water that allow the bugs that cause spoiling to live. Therefore when the outer coating is removed, the nutritional content of the rice is reduced enough to make the food unfit to sustain bugs. It could be argued that it also makes it unfit to sustain humans, particularly when times are hard and rice forms the bulk of the diet, circumstances that prevent the population from satisfying their cravings. The situation with respect to rice is much like being between a rock and a hard place: the people are either hungry because the brown rice has spoiled; risk the detrimental health effects of the toxins that accumulate on badly stored food; or they preempt the problem and process the rice, risking malnutrition. Given that some fungal toxins cause irreversible and fatal damage to the body, malnutrition seems the safer option.

Food processing, packaging and chemical treatments have dramatically reduced human exposure to fungal toxins in industrialised nations, but there is a cost. Preserving rice involves reducing its nutritional content; in many cases the same is true for other foods. The very process of preservation destroys the same fragile nutrients that Hopkins so elegantly proved were essential for life.

As Hopkins was performing his studies in the UK, across the Atlantic the importance of his accessory food factors was being visually demonstrated outside the laboratory, and had been so for more than a decade. In 1912 there was an epidemic disease that was the cause of death for over 40% of citizens; the disease was pellagra and it is attributable to nutritional deficiencies. The symptoms are mouth sores, skin rash and lesions, diarrhoea and ultimately mental problems. We can now diagnose pellagra simply and accurately with molecular medicine, but 100 years ago the diagnosis was based more on what the doctors could observe. In this case, what they could see looked much like leprosy, a contagious condition that prevents pain sensation so that the body becomes damaged and disfigured. Leprosy is also a life sentence for outcast status, social discrimination and, eventually, death. One hundred years ago pellagra was curable, but leprosy was not; however, because of diagnostic limitations, patients with both diseases were grouped together and died together. Fortunately, these diseases can now be distinguished and treated.

The confusion regarding the diagnosis of pellagra makes it difficult to know when it first became common. Certainly it plagued the poor in 1735 when Don Gaspar Casal documented high levels among Spanish peasants. Across the Atlantic, low levels of this disease had been noted in the southern USA for many years. However, in the first decades of the 20th century, crop failures and the economic downturn forced more people on to a maize-based diet, with some molasses and pork fat if they were lucky. Between 1906 and 1940, 3 million people died from pellagra in the USA. As soon as the scourge was declared to be an epidemic, Joseph Goldberg was appointed to find out exactly what this disease was and how it could be stopped. He carried out some studies on inmates of Mississippi prison, who volunteered to participate in exchange for a pardon. This was a farm prison and the men had relatively good health, but when given the diet of maize, molasses and pork fat eaten by the poor people outside the prison walls they soon developed symptoms of pellagra – symptoms that disappeared with supplements of vegetables, meat or milk. With time, Goldberg showed that the cheapest and most effective way to cure pellagra was by adding brewer's yeast to the menu, and yeast-based spreads began to be produced and were popularised as a health food.

Despite knowing which foods cured pellagra, why these foods had an effect remained a mystery for a quarter of a century. In 1937, Nobel Laureate Conrad Elvehjem found that dogs could also develop the disease, and with these models he was able to identify the active ingredient in brewer's yeast. It was niacin, another of the B vitamins. With the chemistry of the cure known, foods could be tested to see how much of this nutrient they contained, and offal, peanuts and wheat germ were added to the food medication list. Peanut butter, which is in many cases formed of mashed, roasted nuts suspended in fat, salted, and sweetened with good old sugar to improve appeal, became added to the "healthy" food list. On the other hand, maize was found to be one of the few cereals that lack niacin and also the amino acid tryptophan, which is important for the synthesis of this B vitamin. Long before modern science was born, Native Americans prevented pellagra by grinding beans with maize to provide the complete set of amino acid building blocks. The same combination of beans and grains is found in staple foods in Mexico, such as corn tortillas and refried beans. As the second stage of nutritional history progressed and people were left to their own devices to choose their diet with less and less interference from Mother Nature, this knowledge was lost, perhaps because beans were made redundant by plentiful meat supplies. When the meat supply ran dry because of famine, it took decades for scientists such as Goldberg to tackle the problem with yeast spreads and peanut butter.

3.3   EARLY CHEMISTS AND NUTRITION

Since the discovery of the citrus fruit cure for scurvy, peanut protection against pellagra and brown rice preventing beriberi, scientists have also shown that a number of other eating plans prevent or alleviate the symptoms of different diseases. Hippocrates was proven right when he stated, millennia before, let food be your medicine. One by one deficiencies were identified and chemists were quick to develop low-cost foods rich in these nutrients and, more recently, supplements in pill form. Systematically, industrialised nations have seen the fatal and severely debilitating deficiencies become confined to history.

Therefore, the third stage of nutritional history saw new food products marketed directly at the new focus on health: yeast spreads and peanut butter, fortified rice and cereal products, and vitamin supplements. However, the health focus was not only on deficiencies but also on the consequences of excess.

By the early 1800s, scientists were able to probe different foods to see what atoms they contained, and what they found was probably quite surprising back then, given the diverse appearance of our meals. All food is made mostly from the same types of atom: carbon, hydrogen, oxygen and nitrogen. Some other atoms are found in different foods; for example, milk is relatively rich in calcium, seafood has more iodine (in the form of iodide) than most foods, bananas are known for their relatively high potassium content and parsley is relatively rich in vanadium. But by far the most abundant atoms that are found in food are the most abundant atoms found in all living things: carbon, hydrogen, oxygen and, to a lesser extent, nitrogen.

A few decades later and the atomic make-up – the empirical formulae – of the three macronutrients, carbohydrates, proteins and fats, was beginning to become clear. Although the atomic arrangement is different, carbohydrates have the same formula as hydrated carbon. Their ratio of atoms is the same as it would be if each carbon atom were attached to a water molecule (two hydrogen atoms and an oxygen). Fats contain the same three types of atom, but with much greater amounts of carbon and hydrogen and much less oxygen. The atomic ratios alter the energy storage potential: the amount of trapped chemical energy is loosely proportional to the amount of oxygen that is added to convert the carbon and hydrogen to carbon dioxide and water, respectively. Therefore carbohydrates, with their higher oxygen content, have less trapped energy than fats. Proteins have an empirical formula more similar to that of carbohydrates than fats, but, in addition, small amounts of nitrogen and the occasional sulfur atom are thrown into their structures. These differences mean that, although proteins can be used to make carbohydrates and fats by stripping back the additional types of atom, the reverse process is much more complicated and protein is an essential dietary component.

In 1842, a top German scientist of the time, Justus Liebig, published his ideas about body chemistry in a book entitled *Die Organische Chemie in Ihrer Anwendung auf Physiologie und Pathologie*, which had such an impact that it was rapidly translated into other languages, including an English version under the title *Animal Chemistry or Organic Chemistry in its Application to Physiology and Pathology*.[3] The book precipitated a wave of recognition of the importance of diet; after all, we are what we eat. Accumulating data allowed what appeared to be a scientific basis on which to build our menus, and soon dietary advice for promoting good health expanded from the enforcement of the sailors' rations of lemon or lime juice to recommendations for health-promoting diets on land. For the sea folk the advice was right; for the land dwellers, it was largely wrong. The Dutch scientist Gerardus Johannes Mulder aired his views on nutrition five years after Liebig's publication. He had studied proteins and discovered their empirical chemical formula and, unsurprisingly, when he came to propose what was good for health, the subject of his studies was given a prominent place. He suggested that 100 g of protein per day for a labourer was a healthy intake.[4,5] He was wrong. The current daily recommended intake of protein is about one-third less, but it was not reduced until it had been raised further. The high protein recommendations were partly based on Liebig's belief that the energy our bodies need to drive our protein-rich muscles must be derived from meat – protein-rich animal muscles. He was wrong. Protein is not the preferred fuel; that is the honour of glucose, and only when the supply of glucose runs short does the body turn to other fuels, such as fat and protein. The mistakes made were understandable, given the infancy of science and technology, but their launch left a lasting impact on the public understanding of nutrition.

Shortly after Liebig swayed the government with his ideas, Edward Smith of Queen's College, London, mounted an opposition. Smith went against the grain of popular belief in science and religion: he opposed leading scientists by suggesting that other fuels could power movement and he opposed the state church with his Wesleyan faith (hence, he could not attend the well reputed Oxbridge institutes of learning because they were at that time only open to members of the Church of England). He proposed about 3000 kilocalories of energy and 81 g of protein per day, which was close to what modern-day scientists would recommend given the lifestyle of the people back then. Whether it was geography or Smith's rebellious nature that led to the dismissal of his intuitive ideas by the powers that be is impossible to say, but no one took much notice of his science and by the turn of the century, the recommended daily allowance of protein was 50% higher than Smith's ideal – perhaps the scientists (along with the public and the politicians) preferred to promote ideas that pandered to their Stone Age taste preferences.

Across the Atlantic, science was approaching nutrition from a different direction: a more simplistic presentation based on food groups rather than hidden components. Wilbur Olin Atwater, the North American agricultural chemist who founded and directed the Office of Experiment Stations, wrote the first dietary guidelines and his ideas were so farsighted they are still used today as the foundation for the food pyramid. His dietary overview emphasised three key points: variety, proportionality and moderation. With the exception of extreme (and often unhealthy) programmes, these ideas are yet to be contested, and yet, because of the infancy of science, he was naïve in some of his ideas. Like Mulder he recommended a huge protein intake and he also strongly advocated use of the calorie as the means to measure the efficiency of a diet. These errors could have contributed to the obesity epidemic. Given the nutritional background in which he was working, his viewpoint is hardly surprising: nutritional deficiencies were epidemic, including energy deficiencies. Academic achievement was beginning to be correlated to diet and schools aimed to improve their statistics, not by cramming children with more information, but by organising and serving a decent meal at lunchtime. "Decent" was defined by the standards of the time, and the thinking was along the same lines as Atwater's – to provide more energy. Indeed, energy had been proven to be essential for decades, whereas the new fad for Hopkins' accessory food factors was only just filtering through.

It is a fact that at the end of the 19th century, and right through to the middle of the 20th century, the average diet lacked energy and adding a little more had visible effects: it helped people to think harder and grow taller. Just as vitamin C supplements tackled the deficiency that caused scurvy, pure calories tackled energy deficiencies and improved health. Sugar not only helped the medicine go down, it was itself medicinal. But as with all medicines, the effect is determined by the dose and there is such a thing as an overdose. Starting a little over 100 years ago, people began to enjoy guilt-free indulgences of their taste preferences and body weight began to climb, but stayed within the healthy BMI range because a trim physique was more desirable than a second portion of food. Most people were able to restrain their appetites enough to keep their waistlines healthy, but there were serious health problems as a consequence of nutrition: heart disease became epidemic long before obesity. Fatty foods had been a central part of the diet for centuries, but now nutritional understanding was eliminating some of the deficiencies and food adulteration had been quenched, people were living long enough for the effects of fat to be fatal. The heart disease problem was in part due to the lifting of the natural food restrictions and allowing free indulgence of genetically programmed food preferences to a health-damaging excess; but the flawed, government-backed nutritional advice must also take some of the blame.

In 1914, Henry Sherman of Columbia University recommended a modest 50% reduction in meat consumption in an attempt to improve heart health. It took a few decades to have any affect. Part of the delay was that the recommendation was a re-education. At the turn of the 20th century, meat consumption was endorsed by the government and now there was a backtracking; the government had given the public permission to indulge their instincts and now they were slapping wrists for the health consequences. The delay was possibly extended because of a blip in the re-education programme caused by the re-launch of low-carbohydrate diets: books were being published that recommended weight loss programmes – visible health improvement – through exclusive eating of the very foods the government now shunned, protein and fat.

3.4 CALORIES DON'T COUNT, AT LEAST NOT PROPORTIONALLY

Historically, the most energy dense foods on the menu are most heavily associated with weight gain. And why shouldn't they be? The energy balance shows that the more calories we take on board through our food, the more calorie-consuming activity is needed to prevent weight gain. This relationship is indisputable. The first law of thermodynamics states that energy cannot be created or destroyed. A calorie is a calorie, they are identical and unchangeable amounts of energy. Yet, in our bodies, the fate of food energy depends on many things: it depends on the way the calorie is stored in the food and also its food component neighbours. How we use calories depends on our diet, our lifestyle, and our genes, and the way these genes are used. These factors bias the energy balance. Such biasing has probably been recognised by many great medical minds throughout history, but was not widely published until the middle of the 19th century. One of the earliest examples in the English language was, rather surprisingly, written by an English undertaker. William Banting was born in 1797 and died when he was a remarkable 81 years young. Like many people of his era, he battled to keep his body weight in check, but unlike many people of his era, for the first 60 years of his life he was losing the fight. He tried all the popular treatments of the time: fasts, spas and exercise regimes, to name but a few. Eventually, he became satisfied with his form after using what is now considered to be a low-carbohydrate approach to dieting. In the last quarter of Banting's life, such diets were beginning to be promoted in Paris for the treatment of diabetes and it is likely that he himself had glucose intolerance. Such an assumption explains his former inability to regulate body weight and the success of the low-carb approach: it has been proven scientifically that reducing the amount of sugars in the diet can help people with glucose intolerance attain a healthy weight, but these benefits do not extend to the wider population.[6] Before his diet, Banting subsisted on bread, pastry, meat, dairy products and beer, with a daily fruit pie and plenty of sugar thrown in. By midlife he was obese. Then he swapped bread for veg, cut out dairy and restricted his sweet food intake to the fruit part of the tart without the pastry and lost weight. His dietary changes were several, and therefore suggesting that his weight loss was due to a lower intake of carbohydrate foods neglects the fact that his new regime added in vitamins and minerals in the form of vegetables. Given that there was no scientific control to validate which of the two changes improved his health, we will never know whether Banting had glucose intolerance or simply malnutrition. In parallel, some of the success of modern low-carb diets may be due to the elimination of some of the unhealthiest foods: fast food, chips and sugary treats. This change alone is likely to have an impact on body weight.

Banting lived a further 20 years on his new diet plan, but he was one of the lucky ones of his era; for most people malnutrition brought them to an earlier grave. At this time, medicine was not as sophisticated as it is today; diagnosis was limited – certainly, prediabetes was difficult to detect other than via the body weight gain it promoted. When mistakes were made in the domain of medical nutrition, they often went in the same direction as our taste preferences, favouring fatty, sweet and high-protein foods. Banting's eating plan was no exception, but he found that his new diet helped keep his body weight under control and was so pleased that he published a letter describing his personal experiences. He soon had many followers. The programme appealed because the diet did not involve hunger and still allowed people to follow their trusted instincts and traditions – feasting on fatty and protein-rich foods – which were in part based on Liebig's errors. Because of the infancy of science, there was no reason to doubt that these favourite foods were not the best.

In addition to following his dietary steps, some people followed Banting's publishing steps. By the 1920s a new wave of low-carb diets were launched into a fertile marketplace: slim had become synonymous with successful and so the number of people discontented with their body weight was rising and the weight loss industry was showing signs of a growth that attracted many entrepreneurs wanting a slice of that cake.

By 1934, the low-carbohydrate dieting craze was evident enough to cause a drop in potato sales in the UK, but this shift in demand was nothing compared to the later bouts. Although body size was important to many people in the 1930s and the following years, rationing and worldwide wars were more prominent on the agenda and size did not become really important until the 1960s when the really, *really* thin look was in vogue. At the beginning of that decade, Herman Taller launched the book *Calories Don't Count*, which recommended a high fat, low-carb diet for weight loss.[7] Taller cited his dieting experiences as evidence that carbohydrate metabolism generated hunger and fat metabolism quenched this effect. He did not consider that his success could be the exception rather than the rule, perhaps the consequence of an underlying medical condition, for example pre-diabetes. Without a controlled scientific study to investigate such a possibility, it will never be known if it was the case. Undeterred, he assigned his success to dietary supplementation with safflower oil, again without a control. Six years later Taller was convicted of conspiracy, mail fraud and Food and Drug law violations for this book; not so much for the recommendations, but because it was believed that the book was produced to sell the safflower oil capsules,[8] and the safflower oil capsules did not meet the claims. According to US law, it was legal for Taller to publish a theory of weight loss and it was legal to market safflower capsules, but only as long as the pills were not mentioned in the book and *vice versa*.[9] The FDA's charges summarised Taller's book as a clever way of promoting the capsules and stated that it was only published for this purpose.[10] However, a significant proportion of the public believed otherwise; just as when Banting launched his letter, some people swore that their weight loss was due to Taller's recommendations. By altering their diet they had managed to bias the energy balance in favour of weight loss – at least temporarily.

On a biochemical scale, there is some truth behind Taller's claims. First, carbohydrate triggers the release of insulin, which promotes fat storage and hunger and, second, fat does trigger the release of satiety hormones. However, insulin release only contributes to weight gain and interferes with weight loss when there are blood sugar spikes caused by quick release from dietary sugars; all carbohydrates are not the same. The safflower oil capsules may slow down digestion and increase the release of hunger-busting hormones, as all good quality fats do; but then again, so does zero-calorie fibre. Fats and sugars both trigger the release of the hunger-promoting hormone galanin, reducing the overall satiety effect, whereas fibre does not. Fats also interfere with the signalling of the hunger-busting hormone leptin. More recently, high-fat diets have been linked to changes in the way the body uses genes to promote metabolic disorders. These changes can be inherited (see Beyond the Helix: Sacrifices for our Children). To be generous, opting for a low-fat, low-carbohydrate diet for a short period could bring insulin levels back into balance and help those with glucose intolerance gain weight control, but there are serious side-effects of long-term low-carbohydrate eating plans.

Health problems arise from trying to fuel the body's functions using fat alone. The final steps of energy release from all forms of food fuels are the same, and involve the citric acid cycle (see The Image of Fat: Body Fat). In chemistry, cycles are nothing to do with bikes, but a series of reactions that begin and end with the same molecule: the citric acid cycle begins and ends with citric acid. However food fuels do not feed in directly as citric acid, rather they feed in as acetyl units, delivered by a co-enzyme that acts as a molecular truck, called co-enzyme A. Despite the common endpoint, the mechanism for getting to that stage is significantly different for different fuels. Carbohydrates are composed of monosaccharide units: chains of six carbon atoms attached to hydrogen and hydroxyl units. In turn, the monosaccharide units may be linked together, in which case the first stage of energy transfer is to release them from each other's grip. The rate of release depends on the number of units in the chain. Granulated sugar has only two monosaccharide units in the molecule, which allows quick release; for starch it takes much longer because there are many units and each must be snipped off in turn. Once released, the journey of the monosaccharide depends on its atomic arrangements. Although several have the same component ratios of atoms, their bonding patterns differ and the exact arrangement gives them different properties in the body and slightly different routes to the citric acid cycle. There are many different types of monosaccharide unit in the carbohydrate family, for example fructose and glucose. Each follows a slightly different journey to become cleaved and processed into a common three-carbon unit called pyruvate. The processing of carbohydrates to pyruvate happens in the absence of oxygen; from then on in, energy release depends on this gas. If pyruvate cannot be processed further because of a lack of oxygen, it builds up in the cell, being temporarily converted to lactic acid; muscle cramps during exercise are a result of this accumulation. As soon as the oxygen supply has caught up with the energy demands, the lactic acid is converted back to pyruvate and channelled into the oxygen-dependent energy release programme: the citric acid cycle.

Carbohydrate processing can take place in any cell that needs fuel, whereas fat processing usually begins in the liver. The basic unit of fats is a fatty acid: chains of carbon atoms from two to thirty or more, adorned with hydrogen atoms along their length and finished with a splash of oxygen to form a carboxylic acid head. The head is responsible for vinegary sourness and has "sticky" properties allowing it to be more easily attached to other molecules. When fats are prepared for the citric acid cycle, the sticky head is used to anchor the molecule and then two carbon units are broken off. For the broken pieces to be attached to the co-enzyme A molecular delivery truck and delivered to the citric acid cycle they need to be modified to have a carboxylic head. So, for each pair of carbons released from the fatty acid chain there are a series of steps to modify them into a suitable form for further energy release.

Just like carbohydrates, if the rate of fat processing for entry into the citric acid cycle is faster than the rate at which this cycle turns there will be an accumulation of intermediates; this time, they are known as ketone bodies. They are produced in the liver and are a normal part of fat metabolism. But when fat is the principal fuel, over-production has some serious health consequences that result in a condition known as ketosis. Like lactic acid, ketone bodies are acidic – sour (confusingly ketone bodies are not part of the ketone chemical family). But unlike lactic acid, which announces its accumulation through the uncomfortable feeling of cramp, ketone bodies accumulate subtly; and they do not stay where they are made but migrate through the body feeding the cells. Dose is important. Normally, the body can cope with a few ketone bodies here and there, but if fats are the only fuel source the body's coping systems become flooded and these chemicals are excreted on the breath and in the urine. The smell is one way to diagnose ketosis and is a sign that the liver and kidneys are under pressure. When both sugars and starches are omitted from the diet and fat becomes the principal fuel, ketosis becomes a probable complication.

It is scientifically proven that over the long term low-carb diets are a high-risk strategy that leads to no more weight loss than a low-fat diet. When dieters are given either low-carbohydrate diets or low-fat diets, the low-carb group loses more weight after six months, but after a year there is usually no difference in body weight.[11] Low-carb diets do seem to have rapid results because glycogen – the muscles' fuel store – is shed along with the water bound up with it, but no more fat is lost than with the healthier low-fat diets. Glycogen loss is a result of the body's dependence on glucose as fuel. Some organs need a constant supply for their functions, and because glucose is not efficiently made from fat, when the diet does not provide this fuel, the shortfall is met by gleaning all that is possible from around the body. The muscles are particularly rich sources because here glucose molecules are tethered together to form a quick-release energy source – glycogen chains – to power the initial burst of movement until fat metabolism kicks in and takes over. Therefore, low-carbohydrate diets see the body tapping into this stock before the fat stores with dramatic, but unsustainable, resulting weight loss. The molecular details of such weight loss are widely known and were the inspiration behind Irwin Stillman's variation on Taller's theme. Stillman's weight loss plan included drinking large amounts of water to convince dieters that fat, not fluid, was being shed.[12] The extra water may give a health boost by flushing out toxins, but will not counteract the weight loss caused by water loss because glycogen is a bit like a sponge, soaking up water and holding it. Without the glycogen sponge, there is nothing to hold the water.

The final nail in the coffin for weight loss via low-carb schemes was provided by Dale Schoeller, a professor of nutritional science at the University of Wisconsin in Madison. In the 1980s, he began to use and promote a nifty way of probing metabolism using water labelled with isotopes of hydrogen and oxygen to trace how these atoms were used by the body.[13–16] Recall that the nature of an atom, whether it is hydrogen or oxygen, is determined by the number of tiny, positively charged particles, called protons, contained in the nucleus (see Why the Fuss about Obesity? Clearing the Name of Chemistry). Hydrogen contains a single proton and oxygen contains eight – never more and never less. In addition, atoms contain neutrons in their nuclei: particles with the same weight as protons, but no charge. Neutrons can be found in different numbers in the same type of atom; normally hydrogen does not contain any, but it can contain one or two. Oxygen normally contains eight neutrons, but can contain one or two more. Atoms with unusual numbers of neutrons in their nuclei are called isotopes. Water made from the uncommon types of hydrogen and oxygen can be tracked in the body. Some of the isotopically labelled water will be lost in urine or on the breath; some of the labelled oxygen will be converted to carbon dioxide during the respiration process; and some will be incorporated into body parts. Comparing high-protein and/or low-carb diets with low-fat schemes, Schoeller concluded that although in the first few months of the programme the people taking the low-carb option did lose more weight than the folk on the calorie-matched low-fat diet, the loss was categorically not due to shrinking fat stores. He postulated that it was due to the loss of fluid.[17]

Since Banting's contribution to weight control, many more derivations of low-carbohydrate diets have been launched and many more will probably follow. The popularity of such schemes does not lie in their success. Rather, low-carb diets may be favoured because they allow indulgence of the genetically programmed food preferences: we are programmed to enjoy protein and fatty foods, and so a diet based around these food choices is destined to be popular; people are more willing to try this approach and more people stick to the plan,[18] regardless of the associated health problems. Some of these health problems are transient, such as the consequences of increased production of small quantities of ketone bodies; larger amounts can cause organ damage. Other effects are more than life-long: eating large amounts of protein and fat, as recommended by some of the more dubious examples of these programmes, alters the way genes are used in the next generation to promote metabolic disorders and obesity (see Beyond the Helix: Sacrifices for our Children).

3.5 MOVING GOAL POSTS

Whilst a foundation of nutritional recommendations was being established, molecular medicine was identifying a new depth of dietary complexity and nutritional needs were changing. When Atwater was preparing what was to become the food pyramid, the average energy intake was lower than ideal and a daily spoonful of sugar had visible effects on growth and intelligence. Unsurprisingly, the calorie became the measure of dietary efficiency and with this measure foods can be easily "improved"; calories are now one of the lowest cost and most readily preserved food components. To score in this goal, simply eat more sugar and more fat. The advice was readily adopted, but as manufacturing began to augment energy availability, needs changed. Lifestyles became increasingly sedentary, with less energy-burning activity to offset what we ate. As people slowed down and less was loaded on to the energy balance to offset the already-in-excess calorie goal, unhealthy weight gain seemed to become a question of time.

Hunter–gathering was once a full-time job for all family members. Children would be taught which foods were good to gather and which should be avoided from early on and would be expected to participate in the daily chores of survival. In the early days of farming, as far as hard work goes, things did not change much. Self-sufficient farmers had to plant seeds, tend to seedlings, raise and fatten beasts, harvest and slaughter, as well as chop, cook, brew and preserve food to make sure that there was an adequate supply for the whole family throughout the year. The toil was so hard that big families lived together to share the burden in a form of labour division. But as division of labour became commercialised and food could be purchased, more and more time was freed to follow other professions. Convenience foods are the ultimate in buying time: compared with home-made meals, they are quicker and often cheaper than preparing foods yourself, when you take into consideration the cost of your time. They free us from the tasks of daily food preparation and shopping. Almost everyone relies on some convenience food or other, be it ready-baked bread, frozen vegetables, chopped meat, potato chips, ice-cream or snack bars – all are more convenient than the basic ingredients used to make the products. It has to be said that in most ways the free time offered by division of labour and convenience foods is good – it has allowed technological and medical developments – but it has also given rise to a new type of problem that is routed in our genetically programmed laziness.

Today, although it is hard to believe, free time is more available than ever. Food gathering takes the form of a trip to the super-market, usually in a car or on public transport, so the only wandering involved is around the various isles therein. Food preparation involves really as much or as little as you want, but there is a limit to the range of basic ingredients that can be bought. Butchery is a skilled job, and most of us would not know where to begin if we were handed a freshly killed rabbit, let alone a cow. The least processed meat products are usually filleted cuts, wrapped in cellophane and often treated with chemicals, such as carbon monoxide, to slow down spoiling. Making a stew simply involves trimming and chopping the meat (unless it is already trimmed and chopped) and tipping it into a pan with a few vegetables. Making this stew is a long way off what our ancestors had to do; even chopping a piece of steak with a flint knife would be a challenge to most of us today, let alone making the knife and cutting the steak off the beast's backside. Flour is ready milled and most vegetables are washed and trimmed. Modern meal preparation, even from the most basic ingredients available in the supermarkets, takes much less time and effort than self-sufficiency; a fraction of the personal energy is used to produce the same meal.

For those who choose, there is another option for feeding yourself that involves hardly any effort at all: ready-made meals. They came about by necessity, the mother of all invention. The necessity was not so much that we the consumers *had* to have these products but rather Gerald Thomas, an executive of C. A. Swanson and sons, *had* to deal with a surplus of 270 tons of turkey after Thanksgiving and he *had* to turn that turkey into profit. Thomas took his inspiration from airline food and gave instructions for scoops of buttered peas and sweet potatoes to be put onto trays with the turkey and packed in boxes that looked like televisions.

He called his product TV dinners. Ten million TV dinners were sold that year – and a new wave in convenience foods was born.

Here was the inverse of self-sufficiency: the complete reliance on other people to prepare the meal from sowing the seeds to putting the food on the plate, and all the consumer had to do was warm it up. Today's world is based on different degrees of dependency and different divisions of labour. Although it is said that division of labour, with each person specialising in a small task repeated over and over again, is the most boring way of working, it is the most efficient. Efficiency saves time. Time is money and saving money in one area allows investment in another to increase the standard of living. TV dinners were successful because they bought time, freeing the 1950s housewives from the kitchen stove; or perhaps convenience meals appealed to a childhood instinct of wanting to be fed by someone else, bringing back memories of "home" when "mom" would cook, serve and present the food and all that was left to do was eat it. These foods also offer more apparent variety and tastes from around the world, which fits in with the idea of an improved standard of living, providing a bit of what is fancied when it is fancied, rather than a menu constrained by the availability and shelf-life of fresh food products coupled to time constraints. The concept is alluring and, with a little help from chemistry, the packages contain edibles that look like home-made sustenance and often taste better: saltier, sweeter and more succulent.

As ready-made meals and other convenience foods have become more popular over the years their price has fallen. New technologies in packaging, flavouring and texturisers have made these products much more appealing, and the relative cost of convenience foods has fallen compared with both the average income and fresh food equivalents. With these changes, genetically programmed laziness and food preferences favour the convenience foods. Their appeal has given them a firm place in society and they have become a bigger and bigger part of our diet. Thus, laziness not only affects health through the exercise not taken, but also through changes in the types of food eaten.

Convenience foods appeal to many aspects of our genetic programming, but in particular our laziness. They save time and effort and, all other things being equal, the easiest option to apparently the same goal is the favoured choice. Fortunately, all other things are not equal and the balance of laziness has to consider the monetary cost of the food too. Even if maths was hard at school the brain is constantly performing subconscious, complicated calculations you know nothing about. These calculations use chemical computing methods and guide the choices you make. They use signallers such as glutamate (an amino acid building block of proteins), balanced with sodium (part of table salt) and also calcium (the mineral of bone-strengthening fame). As we walk around the supermarket we automatically use our mental maths to weigh up the effort involved in food preparation from the most basic ingredients against the cost of buying the food ready prepared. Bread-making is a laborious and skilled job. The 1970s bakery strikes in the UK rudely reminded the public of the skill involved in baking. In the absence of commercial bread products, home-made bread was the order of the day and, instead of the familiar fluffy loaves, out of the oven came hard, inedible blocks. Until domestic, automated breadmakers became widely available few people bothered with the kneading and proving necessary to make bread. The effort involved warrants the price of this product in the shops. In addition to ready-prepared bread being more desirable than a bag of flour and some yeast, many people opt for other convenience foods: frozen prepared vegetables offer more flexible portion sizes, a quick meal compared with the fresh alternatives, and the cost of these products can be less than the fresh food equivalent because the additional cost of processing and preserving foods for long-term storage is balanced by the premium placed on fresh foods to cover spoilage during their storage. This premium can make processed foods cheaper than fresh equivalents simply because a percentage of fresh produce is lost due to spoiling and this cost has to be covered by a premium added to the price of the remaining products sold.

Once food has been processed, its lifetime is extended and further loss is dramatically reduced. The cost benefit extends out of the factory and into the kitchen: if you are trying to limit your shopping to just once a week you will know that you can lose fresh fruit and vegetables to spoiling at home before you have had a chance to eat them. Processing extends the shelf life of food, but also deceptively changes the nutritional balance.

Some foods are subject to a surprising amount of processing. Chicken pieces may be injected with salt-water to improve flavour and plump up the flesh. The effects are significant. Meat processed in this way can be as much as 15% salt-water and this is a pitfall for people on a low-sodium diet. Feed mix ingredients have been shown to affect the colour of egg yolks. Many feeds used in intensive farming include growth enhancers and calories; the growth that is enhanced is usually in the form of fat. Sometimes you can see the difference: the pink flesh of salmon is naturally coloured by the crustaceans eaten out at sea; for farmed fish feed comes in the form of pellets and white fatty deposits can form in the flesh. Modern farming methods have similarly altered the fat profile in other meats. Plant crops do not escape: fruits and vegetables have been selectively bred to be sweeter by removing bitter-tasting phytochemicals and encouraging growth via starch and sugar accumulation. As we are what we eat, is it any surprise that when we eat fatty animals our bodies become fatty too?

Once the food has been harvested, nutritional balance continues to drift. Some nutrients are easily destroyed. Others decay slowly during the storage period because, although the food is safe to eat as packaging and preserving technology protect from bacterial and fungal contamination, there can still be a change in nutritional balance. It is not all bad. The nutritional change during storage alters flavour and was once called maturing. But the nutritional drift for processed foods is different.

Calories are one of the most stable nutrients. When processed foods were first introduced, the population could be considered to be suffering from calorie deficiency and sugar was shown to be a cure. Just as yeast spreads and peanut butter were developed to combat pellagra, fruit concentrates were developed to prevent scurvy and evaporated milk products to prevent rickets, sugary syrups were developed to increase the energy intake. They were used to flavour milk or just water, still or with gas, and in all forms boosted calorie intake, particularly among children. As time progressed, these "medicinal" foods became more numerous and more widely available. Eventually, their "medicinal" effects were forgotten and they became an expected part of the daily diet.

Many food products are now marketed through their health-boosting effects. Take, for example, the vegetarian products that are designed to replace red meat, for example the veggie burger. Often these products require plenty of salt and monosodium glutamate to mimic the meaty flavour, and plenty of fat to mimic the meaty texture. Milk is marketed for its high calcium level, whether it is the natural product or a milk-based drink with plenty of sugar added. A recent billboard campaign claimed that the calories in a cereal bar were fewer than those in an apple; it did not mention phytochemicals and other aspects of nutritional balance. Diet cola is marketed by slim and sexy models, but water is more slimming.

As food products were being developed the nutritional goal posts set in the 1920s were on the move. Technology reduced energy requirements and molecular medicine indicated that the fat-linked health decline began with smaller reserves than once thought; the BMI was revised to define many individuals formerly considered to be of healthy weight as overweight and many formerly overweight individuals as obese. It is currently being revised again; in the same direction. Singapore and Japan are taking a leading part in the reformation. As mentioned before, the BMI is an estimate of obesity based on body weight, but is not a true diagnosis of the disease via body chemistry. People of certain nationalities are biochemically obese before crossing into the international BMI obese zone, and the Japanese and Singaporeans are among them. Given Galbraith's century-old observation that once body weight begins to rise, its control can be considerably more difficult, the leaders of the most vulnerable nations are quick to pass on Lind's two-centuries-old message that nutritional diseases are better pre-vented than cured. The nutritional goal posts moved and food availability changed, both in directions that makes it more difficult to score a healthy body weight.

There are many reports in the media that suggest that part of the obesity problem is a preoccupation with body weight and that the psychological stress caused by being classed as heavy con-tributes to unhealthy body weight. I propose the alternative: the BMI zones do not match the fat chemistry indicators of all nations and those with the greatest mismatch are more at risk of developing obesity because of complacency. Given its northern European foundations, the BMI is most likely to be appropriate in these countries. The people most susceptible to visceral adiposity, for example people with Asian ancestry, are the most vulnerable to unhealthy weight gain. They can be comfortably nestled in the healthy weight BMI zone while their fat chemistry is silently brewing obesity through changes in metabolism and appetite regulation.

As the goal posts for defining daily energy needs narrowed and technology promoted sedentary lifestyles, the demand for other nutrients changed too, and nutritional knowledge widened. By the turn of the millennium science was suggesting that people have different dietary requirements dictated by their genes, lifestyle and health. The recommendations given by the government are averages, not individual statistics. As an example, in the case of obesity, the fat itself necessitates a higher intake of vitamin D and vitamin E: the balance of these same nutrients is involved in weight regulation.

Modern nutrition is not the only factor contributing to declining health. At a molecular level, several key changes in diet combine with the new trend for super-sedentary lifestyles to alter the body chemistry profile and cause a range of health problems, of which obesity is just one. Osteoporosis is another. Richard Klein, palaeoanthropologist and professor of biology and anthropology at Stanford University in the USA, has suggested that "The ever-increasing decline in relative bone strength over the past two million years is most likely explained by advances in technology that have increasingly protected our bodies from physical demands."[19]

Modern living has so many advantages over Stone Age dwelling; in fact, it has many advantages compared to much of human history – better shelter, advanced medicine and super home entertainment systems are just some examples. These advantages are beyond the wildest dreams of our ancestors and so were not the driving force behind progress; for humans there was a more pressing lure – food. And the lure was not just quantity, but quality too. Among the traditionally highlighted reasons for the shift from hunter–gathering to farming and then from farming to food processing, one of the most neglected is the genes that determine food preferences. Both revolutions were driven and directed by this innate programming. As humanity gained increasing power over the food supply, they relied on the same genetically programmed taste preferences that had guided them, and guided them well, in the hunter–gatherer setting. In the natural world, these instincts allowed successful selection of the best diet available. Yet when they were also used to change the availability of food, their health-preserving power was weakened and then lost. The weakening took centuries, but by the 16th century the majority of the British population ate mainly meat and cereal products, and as a result nutritional problems were rife. Those that survived childhood may have lived beyond the average life expectancy of half a century, but, despite having active jobs and eating less than the current recommended daily calorie intake, many battled against unwanted body weight gain, suffered from the consequences of excess fat, salt and sugar and were riddled with nutritional deficiencies. All because they blindly followed their genetically programmed food preferences.

The obesity epidemic could be viewed as more of the same: people continue to indulge genetically programmed greed and laziness, but now that the energy supply has increased and modern living means energy demands are lowered, obesity seems inevitable.

3.6 GLUTTONY

Let's look at gluttony first. One of the strongest forms of evidence that greed has directed the obesity epidemic is that, particularly in the USA, portion sizes[20] and average calorie availability per person[21] have risen with the average girth of the population. The size of a typical 1997 hamburger is six times larger than the 1957 equivalent, the size of a soda accompaniment has increased three-fold in the same period and you would need to order about five 1957-style muffins to meet the 1997 size standards. In the cinema, the size of a serving of popcorn increased more than five times over the same period.[22] At the height of the obesity epidemic, some restaurant portion sizes have grown again: the sizes have gone *super* large, too much for a single person. In parallel, the number of obese people in the USA has dramatically increased.

Links between portion size and girth have been made in Europe, not just by comparing changes in one country over the last few decades, but by comparing different nations right now. The French love food and have no fear of fat or carbohydrate – the nutrients cited as the leading contenders for causing weight gain. Nevertheless, more French people are within the ideal weight bracket than British people. This does not mean that none of the French are obese: there is an obesity problem in France, particularly among the rural populations. Even so, the progression of the disease is not as concerning as in the UK: while the number of obese French people has steadily risen in the past few decades, the number of obese people in the UK has shot up, overtaking many other European countries.[23]

The rapid rise in the incidence of obesity in Britain now puts the people amongst the fattest in Europe, even though not so long ago they were amongst the slimmest. Over the same period, restaurants have become more popular in the UK as the cost of this little luxury falls and there is a heavier reliance on convenience foods.[20] Restaurants, convenience foods and recipe books all give the increasingly plump British population larger meals than eaten by their slimmer French neighbours. Studies show that there is a tendency to eat all the food served in a single sitting[24] (probably associated with culture rather than genetically programmed greed) and, combined, these observations have been a foundation on which to blame greed and portion size growth for epidemics of obesity: portion sizes are larger in the UK than France, and the UK has a bigger obesity problem.

At a restaurant, customers rely on the staff to provide the appropriate portions. Unfortunately, the person doing the portioning does not usually know the nutritional needs of each customer, and often never even meets them. A small person requires a different size of meal to a large person, the average man needs more food than the average woman, and an athlete needs more food than a sedentary friend. One portion size does not suit all.

A note of caution for any young entrepreneur out there who has just had the brainwave of starting a restaurant chain where customers weigh in to receive an appropriately portioned meal: the eyes are involved in determining that feeling of being full because appetite can be transiently regulated by psychological factors, including those that cause dissatisfaction if you receive less on your plate than your luncheon buddy. A reason to grumble, even if it is unfounded, is not conducive to customer loyalty; it is also a reason to have a little snack to fill the space, even if the space is imaginary and nutritional needs have been met. Thus, tailoring meals to meet the needs of each customer is more likely to curb profits than body weight, and so portion sizes look set to rise some more.

With the rise of obesity there has been an escalation in the dependence on people outside the family unit to judge appropriate portion sizes. As far as nutritional history goes, in many countries a complete stranger allocates more of the portions than ever before: eating out is more common and convenience foods form a greater part of the diet. In the USA, one of the countries worst hit by obesity, many folk eat out daily, if not twice daily, and in the amount of pre-portioned convenience foods and snacks consumed the USA is likely to rank as one of the top countries of the world. If each one of these meals provides a larger portion than needed and greed stimulates the diners to eat all the food served, weight gain seems inevitable.

Greed could have initiated increases in portion size – our intrinsic greed may demand more than we need. Such greed can be exploited to boost profit. Often restaurants make more profit on larger, apparently cheaper products than on the smaller alternative. Disproportionate pricing is the direct result of economics. Certain costs associated with meals are fixed regardless of how much each customer eats. For example, it costs a certain amount of money to clean the restaurant and a certain amount of money to rent the building. A premium to cover these fixed costs has to be included in the price of the meal and is usually divided per head rather than per gram of food consumed. If a customer chooses a small portion of food, all the fixed costs still have to be covered just as if they chose a large portion. To simplify the maths, say the fixed costs make up half the price of a small meal, the cost of the food itself is only half the money paid. If the fixed costs stay the same regardless of the purchase, on doubling the food portion, the meal will cost only 50% more than the half portion – an apparent bargain in terms of meal size. The comparatively lower cost of larger meals appeals to our genetically programmed tendencies towards greed and laziness.

The benefits of disproportionate pricing extend beyond restaurant meals to most things that can be bought, including many foods: it is cheaper to buy in bulk. Manufacturers can generate the same amount of profit by selling larger packs at lower cost than the same amount of food distributed into several smaller packages. And making "man-sized" portions is an efficient way to maximise customer satisfaction, even though only half of the population is male. Two different portion sizes involves two appropriately set up production lines, packaging sizes and shipment crates: two sets of fixed costs to be divided over the products sold, and for every additional size option there are additional costs. Therefore, providing the largest portion and allowing smaller and more sedentary people to use their own discretion as to how much they eat is a cost-effective and efficient way of giving customer satisfaction. It just doesn't seem to work for healthy weight control. Perhaps the cost benefits associated with bigger portions are too appealing; if disproportionately more food can be gained from a little more effort, measured in money, this becomes the favoured option for most.

When eating is driven by volume alone, the ultimate in meal value comes with eat-as-much-as-you-like buffets. In such restaurants, the customer is responsible for portioning. However, buffets notoriously fool us into eating more than we need: the variety leads us to try all the foods on offer, and keeping count of the steady trickle of food consumed is more challenging when it is loaded into a single serving. Even Lambert was aware of this trickery and cited single servings as a mechanism he was using to achieve body weight control, although in his case it was not successful. At a single sitting, our eyes have a surprisingly powerful role in gauging what we eat, as demonstrated by the work of, amongst others, Brian Wansink and Jim Painter. Wansink is the head of the Food and Brand Lab at Cornell University. This non-profit unit tries to unravel the behavioural and psychological reasons why people choose to buy and eat different foods. Many of their findings were published in the best-selling book *Mindless Eating: Why We Eat More Than We Think* (2006). Painter is a registered dietician. His "portionsizeme" campaigns are precise and to the point, challenging our tendency to eat too much and exposing restaurants that draw in punters through our tendency to overeat. One such campaign targeted a restaurant that enticed customers to be greedy by promoting an enormous burger and fries free of charge as long as you ate the whole meal. The meal provided close to double the daily recommended calorie allowance for the average person. Painter managed to clear the plate, but had to use his wallet to clear the tab as he correctly portioned the food by dividing it among sixteen friends – the offer provided sixteen times more calories, fat and salt than a dietician would recommend. The portion size he used not only considered energy needs but also other aspects of nutrition: his division left a calorie space for the diners to eat a portion of vegetables and balance the meal. He practised holistic nutrition rather than just gauging needs by calories alone. And his take-home message was a simple taste of French culture: by all means enjoy your favourite foods, but be satisfied with a smaller portion – even a meal targeted at children – that better meets your dietary needs, and then fill up on salad. Yet to implement such control, we need to master our greed.

Wansink and Painter developed an ingenious way of measuring how the visual element of appetite control sways appetite regulation and satisfaction. The dynamic duo designed a restaurant-style table with a difference: in addition to fixing two normal bowls to the table top, two similar, but modified, bowls were added. The modified bowls slowly refilled with soup; so slowly that the diners were unaware of what was happening. Participants eating from the modified soup bowls consumed an enormous 73% more than those pouring their own soup, and after the feeding frenzy, they were unaware of how much they had eaten and did not feel more satisfied than the people in control of their own portion size. The conclusion that "people use their eyes to count calories and not their stomachs" seems logical. Alcohol safety advice has long promoted this message: be in charge of your own glass because when other people top it up, you tend to lose track of how much you drink. And "liquid calories don't register with our appetite controls", as Barry Hopkin, of Chapel Hill's Interdisciplinary Center for Obesity, University of North Carolina, is famously quoted as saying.[25] Liquid calories may be having a significant impact on the body weight of the population.

The obesity epidemic has coincided with cultural changes that centre on food and drinks "to go". A few decades ago, for an adult to eat on the great British streets was a sign of some celebration – or of letting standards slip. It was tolerable for holidaymakers to tuck into fish and chips on the pier or bags of roasted chestnuts in the run up to Christmas, but uncouth for the rest of the population for the rest of the year. Those that needed to drink while they were out usually did so because of addiction, hiding their poisons in paper bags. Now coffee in paper cups seems to be the essential accessory for shoppers, but coffee cup culture brings with it a pitfall for dieters because the calorie content of the cup can vary dramatically. At one end of the scale is the essentially calorie-free black coffee, but top up an espresso with full-fat milk, add a squirt of syrup and top with whipped cream and the calorie content approaches that of a meal – a meal in which the energy supply greatly exceeds the quantity of other nutrients. Such drinks are part of a new wave of energy-dense, nutritionally poor products.

If the idea that the visual element of appetite control regulates energy intake can be extended from soups to solids, it suggests that if our eyes lose track of what we eat, perhaps because they cannot calculate the actual energy value of a fatty morsel, or because of self-refilling soup bowls, energy-dense drinks, free buffet servings or even more abstract distractions, such as stimulating conversations or television dramas watched while we eat, our stomachs may not be able to rise to the challenge of appetite control. Another Wansink and Painter campaign involves plates, not bowls. Over the past decades, they claim, plate sizes have increased. A small plate loaded with a balanced meal can be considered to be full, whereas the same food can be dwarfed on a large plate. And we humans are quick to form habits at meal times such that if we expect our plates to be piled high, we are only satisfied when we see a mountain of food.[26] These habits can be manipulated either way: our eyes can trick us into overeating, but can also register smaller meals as satisfying. Yet there is a debate as to how well plate trickery works;[27] solids are gauged differently from soup. Certainly, the human brain cannot be underestimated in its ability to understand tricks, especially when they involve food. However, it is also important to remember that the visual element of appetite control seems to depend on the period of time considered because even though our eyes regulate appetite at a single sitting, over the longer term fat chemistry kicks in with a much more powerful effect.

3.7 CONTROL

The power of fat chemistry in quenching greed can be demonstrated by the fact that not everyone falls into the trap of overeating even with free servings; there are distinct cultural differences. Whereas the British seem to savour the challenge to eat to excess – eat-as-much-as-you-like, bumper portions and menu options that include more food at an apparent knock-down price – the French are offered smaller portion sizes and fewer à volonté buffets without feeling less satisfied.[28] The French also have less of an obesity problem than the British.

Eat as much as you like restaurants are considered the most economical way of dining in the UK and many other countries, but suggest overindulgence to the point of stupidity to the French. From my own experience, my firm British roots lead me to be surprised about how many French people skip the apparent bargain buffet and opt for a limited portion à-la-carte. The psychology seems to work in reverse: my fellow French diners say they are not hungry enough to get their money's worth, while their English-speaking cousins enjoy the challenge of trying to force down as much as they can – hunger is not a consideration. The slim French are able to control sinful urges, at least at the dinner table.

The even slimmer Swiss exert even more remarkable appetite control. Traditional Swiss meals involve free portions, particularly of fatty foods. Raclette is, fundamentally, melted cheese and potatoes. But rather than the portions being defined by the size of the spud, the meal is cooked at the table using either an open fire, or, more favourably for the firemen, specially designed hot plates to heat the cheese to pouring consistency. The diners eat as much as they want, usually over a period of several hours. Fondue in Switzerland comes in many forms, all involving cheese, oil or chocolate and all with free servings. Even Chinese fondue, which in Asia involves boiling your own greens, meat or shellfish to make a stock base for a noodle soup to close the meal, has been Swissified by the substitution of the water for oil; fondue Chinoise in Switzerland forfeits the noodle soup in exchange for deep fat frying of protein foods that are served with roast potatoes and salad. Yet, despite the fact that Swiss culture centres on buffet-style servings of the most energy-dense foods known to humanity, Switzerland boasts one of the lowest rates of obesity in Europe – a remarkable 8%; just a few per cent above the level of unavoidable obesity caused by genetic programming. What is more, fondue and raclette evenings are often times of overindulgence – a feast that exceeds daily calorie recommendations, but the Swiss resist obesity. They may be gluttons at one meal, but not at the next. This is a critical factor that is often overlooked: trim people – those that fight fat successfully – are able to adjust their appetite over the long term to compensate for a feast and prevent weight gain. They may suffer from moments when they go against their better judgement to be polite – or simply moments of weakness when greed takes over – but they can also tame the urge.

While it is true that people may overeat at a single sitting, people of a healthy weight have protective fat chemistry that is able to regulate appetite control to prevent unhealthy weight gain. It involves several hormone signallers, of which leptin is one. Leptin has several functions, including informing the brain of how much fat is deposited. This signaller is produced in proportion to the fat reserves and, therefore, when the body begins to lay down deposits, more leptin is produced and appetite is quenched accordingly. There are many other signallers and regulators that control the need to feed; more are involved in quenching appetite than firing it up. And appetite suppression is not the only obesity-protection mechanism in place. When rodents are forced on to a high calorie diet they initially overeat with respect to calories. The surge is partially due to the volume of food consumed: they become accustomed to certain portion sizes and when the nosh becomes the calorie-equivalent to a deep fat fry-up, they take time to adjust their visual appetite control. Like trim humans and the majority of the Swiss population, rodents generally do not gain weight unhealthily. Signals follow overindulgence that promote physical activity and increase body temperature to burn the calorie excess.[29] And the more body weight rises, the more mechanisms come into play to prevent it spiralling out of control. In addition to all the usual satiety hormones that quench appetite, the number of leptin receptor switches is increased, maximising the sensitivity of the body to this signal,[30] and reducing food intake over the long term. These same mechanisms are thought to be paralleled in humans: the body has protection against obesity. Even on high-calorie diets, the body is designed to maintain a healthy weight, not just stock-pile resources. This protection is part of fat chemistry – but fat chemistry is failing some people.

The obese have been proven to be less able to compensate for high calorie assaults than their trim peers. When teenagers feast on one fast-food meal, the daily calorie intake for the chubby ones is 20% higher than for the lean ones. The lean teens compensate for the calorie boost through appetite regulation and eat less after-wards.[31] Without such compensation, there is a gain of 44 g (2 ounces) of body fat – not a considerable difference on day one, but after 365 days it amounts to 16 kg (45 pounds) and, even starting at a healthy weight, obesity is the result. These findings shine a new light on Painter and Wansink's soup experiments: the eyes may be involved in judging food intake at a single sitting, but longer term appetite regulation seems to be determined by fat chemistry. It also provides a plausible mechanism to explain how the Swiss can feast on fondue after fondue without an epidemic of obesity: they simply fast the following day. These studies demonstrate how the loss of compensatory appetite regulation is all that is needed to make a person obese.

What exactly is the difference in chemistry that allows some people to gauge their nutritional needs correctly whilst others fail and fall foul to fat? Is the difference simply greed or less controllable functions of the body? Are these functions part of our inheritance, such that those that fail to fast after a feast do so because of their genes? Is the effect – genetic or otherwise – initiated at birth, or does it respond to a specific environmental trigger? And is the difference, be it greed or otherwise, correlated to body weight? It is known that unhealthy body fat makes weight gain more likely. Galbraith recognised this problem in 1916 (see Why the Fuss about Obesity? The Deep Roots of Prejudice). The inability to use compensatory appetite regulation in response to a feast is only found among the obese, which raises the question: does the weakness occur after body weight crosses the BMI barrier into the obese zone or do people wake up one day unable to balance their nutritional needs such that they become obese within a year?

The existence of compensatory appetite regulation mechanisms explains why some people become and stay obese and why some people are protected against unhealthy body weight gain. It also shows how the vulnerability of a nation to obesity can depend on pre-allocated portion sizes: if the obese lose this mechanism of regulation, each overindulgence adds another few ounces of fat to their person and fast food and oversized portions become a danger, whereas when pre-portioned meal sizes are small, those people who cannot compensate for an overindulgence have fewer opportunities to eat in excess. One could propose that the supersized convenience food culture in the UK and USA has allowed the obesity epidemic to hit hard, because once the compensatory appetite regulation mechanism is lost, the generally bigger portions and a heavier reliance on pre-portioned foods leads to overeating. In contrast, those Frenchmen and Frenchwomen who lose their compensatory appetite regulation mechanisms are protected by their culture. The protective effects are not exclusive to the adult population, who are in charge of choosing their meals: in some of the countries with the worst obesity problem, many children eat fast food daily under the guise of school dinners, and some take the practice home with them to receive a double dose. In such an environment, the fate of obesity hangs above everyone, waiting to strike the moment compensatory appetite regulation fails. It is not the oversized portions that cause obesity, but the inability to regulate food intake after such indulgences.

Evidence for the existence of compensatory appetite regulation extends outside scientific laboratories to epidemiological studies of the incidence of obesity. Swiss culture necessitates a high level of compensatory appetite regulation; Swiss people feast and they fast to attain a statistically low incidence of obesity. This culture is evidence that portion size alone does not determine body weight. It is also evidence that overindulgence at a single sitting is not to blame; at least while compensatory appetite regulation is operable. Similarly, even in the countries of the world with the biggest portion sizes, some people remain trim. They may eat fast food and other types of feast but they do not gain significant amounts of body weight because they compensate. However, to feast after changes in fat chemistry have abolished compensatory appetite regulation results in obesity within a year.

What exactly is compensatory appetite regulation? To play Devil's advocate, another term for "a loss of compensatory appetite regulation" could be "greed"; we could say that those vulnerable to obesity are that way because of greed and those that are protected are simply less greedy. The French are less greedy than the British and the North Americans are greedier than the Europeans. And if greed is to blame, then the US and UK portion size hikes could be a simple reflection of greed – greedier people demand bigger portions that their bodies cannot deal with and so they grow fat. As such, we could conclude that the countries with less obesity are less greedy, and that within countries body weight is proportional to greed. All people have it. Some choose to tame it. Others do not. This simple conclusion has a profound impact on prejudice and the future healthcare of the obese. Importantly, if greed is the cause of obesity among the majority of the population, there is no excuse for being outsized.

However, there are problems with equating compensatory appetite regulation with greed. The first is the size and existence of the weight loss industry. If we can control greed at mealtime, would we really go against this same programming and buy into this industry? The second is the rate of change: if obesity is due to greed, people have changed to become greedier over just a few decades. We could also say that the youth of today are greedier than their parents. Among the Swiss, the incidence of obesity is rising with the profile of an imminent epidemic. About one-fifth of adolescents are currently weighing in heavy; many young adults are unhealthily heavier than their middle-aged parents. Countries at the top of the fat polls may dream of attaining such a statistic, but for Switzerland, this rise constitutes a sign that their status as one of the world's slimmest nations is coming to an end.

The link between greed and obesity is further weakened by comparing the calorie availability in different countries around the world with their obesity statistics. The calorie supply per capita per day shows the average amount of food, in terms of energy, the population consumes. It does not differentiate between the amount that enters the body and the amount wasted, which is affected by culture, but it does undermine the role of greed as a cause for obesity. The data span decades:[21] before, during the rise and after obesity was declared epidemic. The pattern is not quite what you would expect if obesity were caused by greed alone. Recall that the order of obesity means that the North Americans top the British who top the French; the incidence of obesity increases as you move west. Logic would predict that the calories supplied per capita would follow the same pattern. In 2003, the most recent data available, an average US citizen was supplied with 134% of the recommended daily calorie intake for the average man and in the UK this percentage was 123. Assuming that availability correlates with consumption, on average each British person eats less than each North American, providing a plausible explanation for the differences in the degree of body fat storage on either side of the Atlantic. Yet, despite smaller portions in restaurants and cookery books, the average French person has a calorie supply midway between the British and the North Americans, breaking the correlation. The anomalies suggest that the French are less greedy than the British in one sense, but more in another: they regulate their appetite to achieve a lower incidence of obesity, but the supply of food suggests that their eyes are bigger than their stomachs and they waste significantly more food than the British. Alternatively, the relative "weight" of a calorie may be less in mainland Europe. The latter proposition sounds the most preposterous, especially given that the first law of thermodynamics states that energy can be transformed from one form to another, but cannot be created or destroyed. Fat chemistry, like other areas of science, is bound by these laws. But there is also a less predictable component: the bio factor. These three letters are derived from the Greek word bios, meaning life, and make up a commonly used prefix that indicates that a word applies to life forms. For example, a hazard becomes a biohazard if it involves live bacteria. In fat chemistry, we can talk about bioavailability of nutrients, which is the amount the body can extract from food. This amount is not necessarily the same as what can be extracted using *in vitro* chemical methods. Our genes, lifestyle, health and other parts of a meal change the amount of nutrition absorbed. The bio factor means that biological systems – living things like us – are not as predictable as mixtures of chemicals in a test-tube and calories do not always count as we would expect. Taller was partially right in his claims: calories do not always count *proportionally*. The bio factor is perhaps best summarised by the great American comedian, William Claude Dukenfield, better known as W. C. Fields, who famously said "never work with children or animals" – the bio factor adds complexity in the field of acting and the same applies to nutrition.

The bio factor can explain the mismatch between calorie availability and body fat. But before considering the explanations, let us consider more cases because the mismatch does not stop at France and the UK: the average Swiss person, living in one of the least obese nations in Europe, has a calorie availability higher than that of the average British person but lower than that of the average French person. In 2003, the Austrians, Portuguese and the US citizens topped the polls for calorie supply amongst the surveyed countries, bathing in more than 132% of the recommended daily calorie allowance for the average man; and at about this time the Austrians and Portuguese claimed obesity statistics around 12.5%, whereas in the USA the figure was more than double.[32] Interestingly, in the same study France reported obesity statistics similar to those of the Austrians and Portuguese, but with a lower calorie consumption, suggesting that the French are not as resistant to obesity as they would like to claim. Now, the calorie availability is based on facts: it can be gauged from product sales. In contrast, there could be errors in the obesity statistics because they reflect what the people report themselves and such figures are usually underestimates. In the USA, both self-reported and measured data are available and when medics measure people in the higher BMI zones the prevalence jumps over the 30% barrier. Despite this potential error, using only our eyes (and a bit of common sense), we can conclude that among these three countries (the USA, Austria and Portugal) the prevalence of obesity is significantly different: the Europeans are significantly slimmer, suggesting some sort of protection against unhealthy body weight gain despite a free flow of calories. The Austrians and Portuguese may waste more food than the North Americans, suggesting that even though they have greater food availability, effective appetite control allows them to eat only what they need, undermining the role of increased portion size as the cause of the obesity epidemic. Their eyes judge portions beyond their stomach's capacity (greed?), but they do not overeat because of compensatory appetite regulation, and so more is wasted. Alternatively, on different sides of the Atlantic the fate of food energy may be significantly different. Energy cannot be created or destroyed, but the bio factor means that it does not necessarily have to become fat. Food energy can be transferred into many different forms, including movement and heat. Fat chemistry determines this fate. The different modes of energy use may be significant enough for Europeans to stay healthier than the North Americans even if they eat the same amount of calories. Importantly, given that fat stores are associated with excess energy, a greater percentage of the people living on the west of the Atlantic seem to be eating in excess. Are they greedier than those living on the other side or have they lost their chemical mechanism of compensatory appetite regulation; do they have an illness that is causing them to overeat?

Looking at the links between calorie availability and the incidence of obesity from another angle gives pretty much the same confusion: examples of countries that boasted over one-fifth of the population as obese in 2005 include, in ascending order, Iceland, Ireland, the UK, Australia, New Zealand, Mexico, and the USA. The calorie supply per capita between these countries does not follow the same pattern. For example, the Australians choose to munch on 4% less than the Icelanders, but have an incidence of obesity 4% greater. Climate! I hear the smart ones cry. But, within reason, the climate in New Zealand and the UK is not that dissimilar, yet the Brits have 8% more calories at their disposal and a lower incidence of obesity. The poor Mexicans have a lower calorie supply than the British – a whole 10% lower – and yet they are riddled with a 25% higher incidence of obesity compared with the Brits. Once again, either the differences are due to waste, indicating that people in the trim countries have superior compensatory appetite regulation and take only what they need from a substantial excess, or the amount of waste is the same and people in the countries that eat more stay slim because of the fate of those additional calories. Either way, those with the most calories available are not the fattest people, which lays greed to rest as a cause of obesity.

3.8   INTRODUCING LAZINESS

The energy balance has two sides: one to weigh calories and the other where this load is counterbalanced through calorie-burning activity, for example taking a walk or shivering in the cold, using metal blocks with weights proportional to the calories used in these activities. If the balance tips to the eating side, we gain weight, and if the balance tips to the energy burning side, we lose weight. Using this pictorial representation of the energy balance, it is easy to see how *either* eating all we want *or* doing as little as we can get away with can amount to an energy surplus that is stored as fat. This balance illustrates how greed alone may not cause obesity, yet combined with laziness, weight gain can result. The Swiss may not be superior to the British when it comes to controlling their greed, rather they may be more active.

Technology has transformed our lives, reducing the need to work manually, and even aiding with many of the traditional tasks around the house. Modern-day technology makes survival jobs significantly easier compared with in times gone by, and also facilitates many of the jobs we prefer to be done, but are not crucial (such as vacuuming). It is not necessary to defend our territory more than maybe trimming the hedge or painting the fence once a year. Annual repairs to shelters are minimal owing to better construction methods and structures that are more durable. Few people make their own clothes, let alone cooking utensils and tools. Now, if left unchecked, genetically programmed laziness could promote super-sedentary lifestyles: technically, people can survive without leaving their beds, even if they are not rich enough to buy slaves. This lifestyle is tried and tested by the record-holding big guys and gals. To obtain the title of the fattest person alive in the *Guinness Book of World Records* one has to beat people who are completely immobilised by blubber, such as Manuel Uribe.

Uribe's lifestyle is an extreme, but part of a spectrum that touches all our lives. The current imbalance between diet and our genetically programmed laziness, pushing us towards sedentary lifestyles, is a logical and attractive explanation for the obesity epidemic and one that seems to be supported, at least in part, by scientific studies. For example, over a five-year period in the 1980s, a surveyed section of the Dutch population improved their diet to include more vitamins, less cholesterol and fat and, overall, fewer calories.[33] However, although the dietary change improved general health, it had little effect on obesity because the Dutch became more sedentary in this same period: they ate less, but they also did less and, therefore, their body weight did not change.

The survey mentioned above – part of the Dutch National Food Consumption Survey (DNFCS) – demonstrates just how detrimental lazy lives can be to weight control over a five-year period. Over the last 50 years, increasingly sedentary lifestyles have reduced the average energy expenditure in the UK by the equivalent of running a marathon each week;[34] that's a lot of blocks! At the beginning of the third stage of nutritional history, energy was in short supply; by its end, enough energy-dense products had been launched onto the market to turn a deficiency into an excess. Throughout this period, lifestyle became increasingly deskbound. Now inactivity has reached such a level that one has to ask how super-sedentary folk can reduce their energy intake sufficiently to prevent weight gain without causing deficiencies of other nutrients. For these people, a standard sized fast-food meal may provide more calories than needed in a day, without satisfying vitamin and mineral needs. And a supplement in pill form does not cover all bases. Could this imbalance be perturbing appetite regulation? After all, if compensatory appetite regulation is fully functional, it should work independently of activity levels, adjusting appetite to meet needs without weight gain. The link between lazy lives and unhealthy body weight gain becomes a function of failing compensatory appetite regulation.

3.9 MORE POWERFUL THAN THE CALORIES WE CONSUME

The changes in physical activity over the last few decades are a consequence of all-round technological developments. Examples include less physically demanding employment, upgrading the bicycle for the motor car and, one of the favourites for shouldering the blame for the heavy burden of obesity, television watching. Part of the focus on TV viewing as a cause for obesity is the easy availability of statistics – TV is such a powerful marketing tool that much money is invested in researching how to optimise its use. The research includes probing how much TV people watch and how they do it. The general conclusion from more than one source is that the hours spent watching TV are steadily rising along with the time spent on other super-sedentary pastimes; television watching, internet browsing and computer games are rapidly replacing more energetic activities. These changes have taken place as obesity has gripped us, suggesting a link.

A survey of over 4000 American children aged between eight and sixteen years showed that children watching four or more hours of television each day were statistically fatter than those who watched less than two hours.[35] The group of researchers who conducted this survey on childhood obesity were, rather surprisingly, from the Department of Geriatric Medicine and Gerontology at John Hopkins Medical School, Baltimore, USA. Usually, this department focuses on degenerative diseases and frailty, but the increasing incidence of obesity in all sectors of the population caused them to look more widely for a solution, in particular focusing on the roots of the problem: fat children tend to become fat adults. They took the data from the National Health and Nutrition Examination Survey III and discovered not only the link between fatness and sitting in front of the box but also that vigorous exercise was on the decline. TV watching, one of the most sedentary pastimes, seems to be replacing sporting activities and the pictorial energy balance demonstrates how this change alone could cause obesity. Yet, just as we should be able to compensate for a feast through appetite regulation, one would imagine that we should also be able to compensate for a lazy day.

In science, the exceptions that do not fit the rule often help the understanding of a problem as much as establishing a rule.

The National Health and Nutrition Examination Survey III
established a general rule: TV watching is linked to obesity. This
generalisation may be extrapolated to suggest that laziness makes
us fat. However, some significant exceptions to this rule were
unveiled by the 2001–2002 Health Behaviour in School-Aged
Children Study. These exceptions were published by a Canadian
group based at the Department of Community Health and Epidemiology, Queen's University, Kingston, Ontario and demonstrate the importance of considering the whole energy balance and
not just one side of it. Unlike the first survey, which limited data
collection to one continent, the 2001–2002 version went worldwide
and involved over 30 times more children. Physical activity was
generally lower and television viewing *generally* higher in the
countries with the highest incidences of obesity; it is clear that
sedentary lifestyles do contribute to unhealthy body weight gain.
But there were some interesting exceptions: of the 34 countries
analysed, the Maltese and North Americans weighed in heaviest,
with over 30% of the surveyed population being obese, and the
Lithuanians and Latvians were the leanest, boasting less than 7%
in this unhealthily high body weight category.[36] However, the
citizens of both lean countries claimed to partake in more TV
watching hours per week than the people living in the USA.

In order for the Latvian and Lithuanian TV addicts to stay trim
when their Maltese and North American cousins pile on the
pounds, the L-countries must have a secret way of achieving energy
balance. Given the cultural diversity in the USA, with many
neighbouring states having equivalent differences to neighbouring
European countries, rather than trying to delineate the integrated
average of the United States of America perhaps it is more pro-
ductive to restrict comparisons to within Europe to identify differences that could protect some people – but not all – from obesity.

Let us start by using the energy balance as we know it to explain
the different prevalence of obesity in the Maltese and the Balts, as
the Latvians and Lithuanians are collectively known. The Balts
have amongst the highest rates of TV watching in the world and
many devote their free time to other sedentary activities.[37] However, calorie expenditure is cumulative throughout the day and the
Balts have the tendency to work long hours outdoors. Eighty
percent are involved in active jobs,[37] for example construction,
forestry, carpentry and, to a lesser extent, the catering industry.[38]

In contrast, Malta's principal industry is tourism; many job descriptions match those in northern Europe – catering and chambermaids – but a number of more sedentary professions, including desk jobs and taxi drivers, are as common.

A lack of exercise has been highlighted as a major problem in Malta, as it has in other cultures facing unhealthily high body fat problems.[39] In the 1980s, the Maltese government launched an urgent package to deal with this problem. The recommendations were quite conventional, suggesting a reduced calorie load and a greater attempt to counterbalance it with exercise. The urgency of the package was because of the bleak outlook for the Nation's health, particularly as the incidence of childhood obesity increased, but also because of the enormity of the estimated healthcare costs associated with the low fitness of the Maltese when considered against the landmass of this island, its population and its gross national product. Ideas and schemes varied and targeted all age groups, including encouraging children to participate, for example, in walking buses to and from schools.[40]

The campaigns have had some success, in part because not only the Government was involved. The poor prognosis for the future of the Maltese shocked the famous into action, including one of Malta's national heroes, Pippo Psaila. His achievements include managing Malta's national football team to achieve their first World Cup qualifying match. He went on to become the Director of Sport in the Malta Olympic Committee and directed the Games of Small States of Europe when it was hosted in his homelands. So it is hardly surprising that when he turned his attention to politics, he applied his interest in sport to tackle the obesity problem. He proposed to reform sport in schools by providing more facilities. However, even when endorsed by such a respected star, more facilities are not reason enough to improve fitness; people have to use the new facilities to become fit. The National Audit Office discovered that the Maltese children were not fond of physical exercise. Maltese sports teachers described their students as showing "apathy and reluctance" towards any exertion. But the lethargy could be a result rather than a cause of the weight gain because being heavy reduces the inclination towards exercise. Still, such comments make sloth seem like a strong candidate for the origin of obesity, except . . . .

When sloth is measured by the amount of TV watching, there is a general correlation with obesity, but there are also some exceptions; similarly, when sloth is measured by a lack of physical activity, there is a general correlation with obesity, but there are also exceptions. In a survey of 33 countries, 29 could be used to correlate excessive body weight inversely to physical activity; four were exceptions to the rule. Just as the Latvians and Lithuanians provide the TV-watching exception, the Maltese provide the exercise exception. Despite the reluctance in the classroom, the Maltese do participate in sport, partly because of Psaila's promotion. The Maltese exercise slightly more than the French, where obesity is much less of a problem, and the French, on average, are supplied with more calories per capita than the Maltese. To put it another way: on paper, the French are greedier and lazier than the Maltese, but are less likely to suffer from obesity. With such data we can only conclude that the pictorial energy balance gives a general relationship between gluttony, sloth and body weight, but in some countries of the world, such as Malta, it is biased in favour of weight gain whereas in other countries, such as France, it is biased in favour of a more healthy weight, and the bias is due to fat chemistry.

The differences in calorie supply and obesity between Latvia and Lithuania illustrate just how far fat chemistry can bias the energy balance. The average Latvian is supplied with 7.6% more energy than the daily recommended amount for an average man; about the amount in a bagel. Now, this value is the supply rather than the amount consumed and is an average, some may consume much more and some much less, and it is averaged over the whole population – men and women – despite the difference in the recommended calorie allowances between the sexes. Therefore, a significant percentage of the Latvian population has the potential to overeat, and overeat to an obesity-promoting extent, but they seem to defy weight gain.

Compare Latvia with obesity-strained Malta, where the excess calorie supply jumps to 26% – about the same number of calories as in a large portion of fries. Wow! Most people do not need hard facts and figures to accept that such an excess leads to unhealthy body weight gain. Based on the pictorial energy balance it is easy to imagine how such a diet could lead to an obesity epidemic and the picture has "greed and sloth" stamped all over it. But, before the case is closed, consider Lithuania. Here obesity is as low as in

Latvia, but the calorie supply is 20% more than what health officials recommend; a medium portion of fries. Can choosing medium really make such a difference to body weight when both sizes are in excess of needs? Common sense says not and science seconds that idea: if the energy cannot be destroyed, it has to go somewhere. Either the eyes of Latvians and Lithuanians are bigger than their stomachs and although they take an excess (greed) they waste more because of fat chemistry-based appetite regulation. Or calories don't count proportionally.

Malta is what you would expect: the Maltese overeat considerably and suffer for it. Latvia you could explain with a bit of juggling: we know that the body has mechanisms to protect against body weight gain – increasing metabolic rate and tendencies towards exercise – such that an additional slice of bread with a meal could be wasted by these adaptations. The case of the Lithuanians, who have genes and a lifestyle very similar to those of the Latvians yet calorie consumption closer to that of the Maltese, suggests that fat chemistry can bias the energy balance. It reveals how unfair modern obesity can be: some people eat less than their neighbours, in terms of calories, exercise more but cruelly suffer from obesity-related complaints, whilst others enjoy a superior level of health with no extra investment. By understanding how the energy balance can be biased we can protect ourselves from obesity.

Throughout the third stage of nutritional development, humans have been left pretty much to their own devices to develop food products that match nutritional requirements. The requirements reflected both needs and preferences. The involvement of chemistry accelerated and streamlined the food production process. At the turn of the 20th century, calories were found to be deficient and supplements of sugar had a remarkable effect on growth and development. Over the following decades, other nutritional deficiencies were identified. Food production met these needs with supplementation and fortification. With these advances, the incidence of debilitating deficiencies has fallen dramatically. But almost as soon as nutritional needs were defined, they began to drift because of changes in diet and lifestyle. Calorie requirements have fallen, but the need for some other types of nutrition is greater than before. Food production has responded by marketing a series of low-calorie products side-by-side with super-supplements supposed to improve health. Yet around the world health is declining, particularly because of body fat. Key findings show that the obese are unable to alter their appetites to buffer a calorie excess, but not because of greed, because many of the countries with the highest food supply have the leanest populations. Similarly, laziness does not account for the different prevalence of this disease around the world. Rather, it seems that obesity has become a problem because of energy balance-biasing. Throughout most of British history, people have battled with body weight and won. They worked hard, ate less than the current recommended daily calorie intake, but still found it necessary to watch their weight; the energy balance was biased in favour of weight gain. In contrast, some people of the world eat one-third more energy than nutritionists would recommend, enjoy hours of sedentary activities and do not become fat; their energy balance is biased in favour of a healthy weight. The epidemiology of obesity shows just how far fat chemistry can bias the energy balance to prevent unhealthy weight gain by directing energy transfer along a path that offers protection. It also suggests that obesity is the result when it fails.

REFERENCES

1. P. J. Aggett, J. Bresson, F. Haschke, O. Hernell, B. Koletzko, H. N. Lafeber, K. F. Michaelsen, J. Micheli, A. Ormisson, J. Rey, J. S. de Sousa and L. Weaver, *J. Pediatr. Gastroentrol. Nutr.*, 1997, **25**, 236–241.
2. F. G. Hopkins, *J. Physio.(Lond.)*, 1912, **44**, 425–460.
3. J. Liebig (W. Gregory, translator), *Animal Chemistry or Organic Chemistry in its Application to Physiology and Pathology*, Owen, Cambridge, MA, 1842.
4. E. Todhunter, *J. Am. Diet Assoc.*, 1954, **30**, 1253.
5. E. Todhunter, *J. Amer. Diet Assoc.*, 1961, **39**, 475.
6. A. Pittas, S. Das, C. Hajduk, J. Golden, E. Saltzman, P. Stark, A. Greenberg and S. Roberts, *Diabetes Care*, 2005, **28**, 2939–2941.
7. H. Taller, *Calories Don't Count*, Simon & Schuster Inc., New York, 1961.
8. Twelve-calorie count, *Time Magazine US*, Time Inc, New York City, 1967.
9. http//www.trivia-library.com/c/history-and-benefit-of-diets-calories-dont-count.htm (accessed 10 May 2012).

10. Calories do Count, *Time Magazine US*, Time Inc., New York City, 1962.
11. A. J. Nordmann, A. Nordmann, M. Briel, U. Keller, W. S. Yancy, B. J. Brehm and H. C. Bucher, *Arch. Intern. Med.*, 2006, **166**, 285–293.
12. I. M. Stillman, *The Doctor's Quick Weight Loss Diet*, Pan, 1969.
13. D. A. Schoeller, *J. Nutr.*, 1988, **118**, 1278–1289.
14. D. A. Schoeller, L. L. Levitsky, L. G. Bandini, W. W. Dietz and A. Walczak, *Metabolism*, 1988, **37**, 115–120.
15. A. M. Prentice, A. E. Black, W. A. Coward, H. L. Davies, G. R. Goldberg, P. R. Murgatroyd, J. Ashford, M. Sawyer and R. G. Whitehead, *Br. Med. J.*, 1986, **292**, 983–987.
16. L. G. Bandini, D. A. Schoeller, V. R. Young and W. H. Dietz, *Am. J. Clin. Nutr.*, 1987, **45**, 868.
17. A. C. Buchholz and D. A. Schoeller, *Am. J. Clin. Nutr.*, 2004, **79**, 899–906.
18. W. S. Yancy, M. K. Olsen, J. R. Guyton, R. P. Bakst and E. C. Westman, *Ann. Intern. Med.*, 2004, **140**, 769–777.
19. R. Klein, *The Human Career*, University of Chicago Press, Chicago, 1999.
20. J. H. Ledikwe, J. A. Ello-Martin and B. J. Rolls, *J. Nutr.*, 2005, **135**, 905–909.
21. *EarthTrends: Nutrition: Calorie supply per capita*, http://earthtrends.wri.org/searchable_db/index.php?theme=8&variable_ID=212&action=select_countries (accessed 18 October 2009).
22. J. Putnam, *Business Library*, CBS Interactive, 1999, **22**, 2–12.
23. R. Jackson-Leach and T. Lobstein, *Int. J. Pediatr. Obes.*, 2006, **1**, 26–32.
24. D. A. Levitsky and T. Youn, *J. Nutr.*, 2004, **134**, 2546–2549.
25. B. Popkin, C. Albala, S. Benjelloun, L. Bourne, G. Cannon, D. Coitinho, C. Doak, O. M. Galal, H. Ghassemi, G. Harrison, V. Kosulwat, M. J. Lee, T. Maletnlema, V. Matsudo, C. Monteiro, M. I. Noor, K. S. Reddy, J. Rivera, A. Rodriguez-Ojea, R. Uauy, H. H. Vorster and F. Y. Zhai, *Public Health Nutr.*, 2002, **5**, 279–280.
26. L. H. Epstein, K. A. Carr, M. D. Cavanaugh, R. A. Paluch and M. E. Bouton, *Am. J. Clin. Nutr.*, 2011, **94**, 371–376.
27. B. J. Rolls, L. S. Roe, K. H. Halverson and J. S. Meengs, *Appetite*, 2007, **49**, 652–660.

28. P. Rozin, *J. Nutr. Educ. Behav.*, 2005, **37**, S107–S112.
29. A. A. Butler, D. L. Marks, W. Fan, C. M. Kuhn, M. Bartolome and R. D. Cone, *Nature Neurosci.*, 2001, **4**, 605–611.
30. S. Lin, L. H. Storlien and X. F. Huang, *Brain Res.*, 2000, **875**, 89–95.
31. C. B. Ebbeling, K. B. Sinclair, M. A. Pereira, E. Garcia-Lago, H. A. Feldman and D. S. Ludwig, *J. Am. Med. Assoc.*, 2004, **291**, 2828–2833.
32. http://www.oecd.org/document/16/0,2340,en_2649_34631_2085200_1_1_1_1,00.html (accessed 18 February 2011).
33. M. R. H. Lowik, K. Hulshof, L. J. M. van der Heijden, J. H. Brussaard, J. Burema, C. Kistemaker and P. J. F. de Vries, *Int. J. Food Sci. Nutr.*, 1998, **49**, S5–S68.
34. *Tackling Obesity in England (House of Commons Papers)*, Stationery Office, Great Britain, 2001.
35. R. E. Andersen, C. Crespo, S. J. Bartlett and S. J. Pratt, *J. Am. Med. Assoc.*, 1998, **280**, 1231–1232.
36. I. Janssen, P. T. Katzmarzyk, W. F. Boyce, C. Vereecken, C. Mulvihill, C. Roberts, C. Currie and W. Pickett, *Obesity Rev.*, 2005, **6**, 123–132.
37. J. Pomerleau, M. McKee, A. Robertson, S. Vaasc, K. Kadziauskiene, A. Abaravicius, R. Bartkeviciute, I. Pudule and D. Grinberga, *Prev. Med.*, 2000, **31**, 665–672.
38. M. Hazans, *Latvia: Working Too Hard?*, University of Latvia and BICEPS, Latvia, 2005.
39. J. Esparza, C. Fox, I. T. Harper, P. H. Bennett, L. O. Schulz, M. E. Valencia and E. Ravussin, *Int. J. Obes.*, 2000, **24**, 55–59.
40. T. Lobstein and H. N. F. Branca, *The Challenge of Obesity in the WHO European Region and the Strategies for Response Summary*, WHO Regional Office for Europe, Geneva, 2006.

CHAPTER 4

Stone Age Obesity

Historically, fat chemistry's failures were rare, but when they did occur it was with a force that could break down the strongest of Mother Nature's barriers. Consider the environment where unhealthy weight gain is least expected: amongst the hunter–gatherers. There are no well-publicised modern cases. Still, solid evidence for ancient obesity comes from the Venus figurines unearthed across Europe. These treasures depict obesity in the human form and some are thought to date back to the Stone Age. The Venus of Willendorf is a particularly fine example. This 11 cm statue was carved out of limestone apparently back in the Palaeolithic period, around the beginning of the Stone Age. Its purpose is one of speculation but a fertility talisman, made voluptuous because of the swollen nature of pregnancy, is a hot favourite. However, rather than increasing the chance of conception, modern medical records categorically show that blubber blocks reproductive potential, and it is naïve to assume that the Stone Age people were not aware of this fact. As an alternative to fertility, these sculptures could have been talismans thought to maintain youthfulness. Large fat deposits under the skin give a baby-like sheen and such links are still made in modern society; consider Buddha, a model of mental youthfulness and obesity in one. Other people have proposed that the Venus figures are evidence that the Stone Agers favoured the obese stature and this statue was for their titillation. Equally valid is the possibility that

Fat Chemistry: The Science behind Obesity
Claire S. Allardyce
© Claire S. Allardyce 2012
Published by the Royal Society of Chemistry, www.rsc.org

these carvings were a type of medical record – sculptures of a spectacle too bizarre to describe. Curiosity funded the North American tour of a wax model of Lambert decades after his death and, back in 1810, curiosity funded an exhibition in London of the Hotentot Venus – who was actually a live woman taken from South Africa.[1] Is it that unreasonable to assume that curiosity could have funded a smaller, more ethical Stone Age equivalent? Whatever the reason for their creation, the fine details of the Venus figures suggest that they were modelled on real women; the bangles and bracelets may be the result of artistic licence, but the fat rolls fold as they would on a truly obese body. Some figurines have cascading layers of blubber familiar amongst modern examples of obesity; in others the fat distribution is characteristic of steatopygia, a genetically predetermined body shape where fat is store disproportionately around the buttocks and thighs; a predisposition most commonly found among some indigenous people from southern Africa. The Venus figurines, however, are most commonly found between the Pyrenees and the plains of Siberia. There are other interesting features of the figurines that suggest they were modelled on real women. For example, the Venus of Willendorf has *genu valgum* (knock knees). Although in the modern day such physical deformities are characteristic of the strain caused by excess body weight, in the Stone Age rickets caused the problem to be more widespread. Such attention to detail suggests that the Venus figurines were a meant to be a true likeness of their models – Stone Age medical marvels, the Yagers and Lamberts of that era.

During the Stone Age, obesity is likely to have been very rare. Most hunter–gatherers and early farmers were probably underweight, as was Ötzi the Iceman. His BMI was $18 \, \mathrm{kg \, m^{-2}}$ and there is no reason to presume he was anything but average. For the majority of the Stone Age population survival eliminated the possibility of growing fat, even if one loved food or liked to laze. In such situations, the novelty of body fat may have given it added value and it may have become a desirable must-have. Kings and nobility may have fed themselves (or, given the associated health problems, just their wives) to the obese state to display their wealth. And yet too much fat was considered a medical problem. As soon as medical records began to be made, obesity featured. Back in the 1st century, when the famous Greek physician Galen was in Rome he noted a patient named Nicomachus of Smyrna, who was bedridden by his blubber until "Asclepius [the ancient Greek god of medicine] healed him".[2] Then there was Dionysius of Heraclea, a tyrant ruling over the region from about 336 BC, who ate so much that his fat not only stopped him from moving but also intermittently stopped him from breathing whilst he slept: severe sleep apnoea. And so he ordered his loyal subjects to prick him should he fail to respire. Eventually his blubber prevented him from eating, and so he was fed "artificially" (whatever that means). After 30 years of tyranny, his own flesh choked him and he met his maker. Such examples of obesity among the wealthy suggest that their resources lifted the traditional barrier to food availability and their greed was left to its own devices. However, the Stone Age carvings do not depict a state of pride, as would be expected for kings and senators. The woman represented by the Venus of Willendorf does not have the poise of a princess or a goddess; she is hanging her head so that her face cannot be seen. The visible part of the head is decorated with carvings that could be braided hair; the texture has also been proposed to be some sort of basket woven to cover her identity. Similarly, the focus of the other Venus figurines is on the fat and not the face.

Outside the ranks of the nobility, genetically predestined obesity afflicts a handful of people. Inheritance can determine body weight and there are many variations within genes that promote heavier or lighter frames within a healthy range, which is why we have a healthy weight range rather than one healthy weight. But some varieties of genes go beyond health. They make hunger an overwhelming compulsion, a drive to eat that stimulates foraging even when other people are content. Back in the Stone Age, this behaviour was unexplainable, even intriguing, perhaps so intriguing that the behaviour, rather than the physical shape, was the inspiration behind the Venus figurines. Genetically predestined obesity can involve a dedication to feeding that goes above and beyond normal satiety levels and overcomes fatigue. Picture the tribe taking their well earned and essential rest, perhaps relaxing around the fire in the evening, while watching, without understanding, their genetically predestined obese cousin as he or she continues to risk all and frantically forages into the night. Such cases of obesity have been documented in modern times and technology now allows us to understand this type of foraging on a molecular level. One of the first cases of morbid obesity to receive the molecular treatment involved two cousins, both with an inherited fat chemistry profile that did not allow them to produce the body fat-gauging hormone leptin. This signaller communicates when our fat reserves are full, sending its message to various control centres so that appetite and metabolism can be regulated accordingly. The cousins became morbidly obese in their early years because they overate – in fact they ate constantly. They are described to have panicked when food was not in view. They were being suffocated by fat, but their fat chemistry said they were starving and so they behaved as though they were starving and went to extreme lengths to eat.

The cousins were living examples of the power of genes in appetite control. Their inheritance precipitated a change in their fat chemistry that predestined chronic hunger and a drive to overeat that was so powerful it overcame dietary restrictions imposed by parents and doctors; it probably could have overcome Stone Age food restrictions too.

It is worth taking a moment to consider these powerful genetic pre-determinants of obesity from both angles: when leptin signalling is abolished, a person will rapidly eat themselves to obesity and, therefore, leptin is a signaller that prevents frantic foraging into the night. It protects against unhealthy body weight. It is proof that the body is not programmed towards obesity. Leptin is not alone in offering this type of protection; there is a whole battery of body fat monitors, appetite control signallers and metabolic valves that protect against unhealthy body weight gain. Indeed, there are actually more signallers involved in suppressing appetite than in rousing the need to feed.

4.1 WHAT ARE GENES?

Our understanding of the causes of obesity has benefited from up to date technology and taking a closer look at genes. At a molecular level, genes are string-like molecules made up of sequences of four different bead-like units named thymine (T), adenine (A), guanine (G) and cytosine (C). Collectively these beads are known as deoxyribose nucleotides in the test tube and when strung form deoxyribose nucleic acid (DNA). All of the approximately 30 000 human genes, like the genes of most living things, are built of

DNA. The genes of mice and men, as well as those of other animals and plants, are strung one after another interspersed with other regions of DNA to make chromosomes, of which we humans have 23 pairs. Together, all the chromosomes make up our genome.

The widely accepted scientific definition of a gene is "a locatable region of genomic sequence corresponding to a unit of inheritance, which is associated with regulatory regions, transcribed regions and other functional sequence regions."[3] A translation of this could read as follows: "a gene is a discrete region of DNA which contains recipe information in a code form. This information is a message that contributes to making us what we are. Reading this message is controlled by other regions of our DNA, some of which are genes and some are not." The control regions of DNA are much like an index in a recipe book, guiding the DNA reading machinery to the appropriate place on the chromosome. Regardless of whether the genes have direct functions or are involved in the regulation of the use of other genes, their sequence is important. They encode instructions for when and how to make proteins and functional ribose nucleic acid (RNA) molecules, and when these units are made, they affect body chemistry.

Like DNA, RNA and proteins are also formed from strings, but using different sorts of beads. The RNA bead building blocks are similar to those of DNA, but there are differences. First, the string part of RNA, its backbone, has an additional oxygen atom tethered in place. In fact, the name DNA states that it is RNA with an oxygen atom removed; it is deoxygenated RNA. In addition, one of the actual beads – one of the four information-coding parts – of RNA differs from that of DNA. As mentioned above, DNA is composed of sequences of T, A, G and C, whereas, in the RNA sequence, T is replaced by uracil (U). U is chemically similar to T apart from a single hydrogen atom replacing a more bulky methyl group (a carbon tethered to three hydrogen atoms) on the structure.

In contrast to DNA and RNA, proteins are composed of structurally distinct building blocks, of which there is much more variety. Proteins are made of sequences of 20 amino acids. Further variety can be obtained by tethering small parts of other molecules onto these basic units, for example, carbohydrate or fat units. Yet proteins are still polymers, like DNA and RNA. The backbone of the repeating units is more flexible than that of nucleic acids, allowing protein chains more freedom to coil and bend, twisting around themselves and other molecules to form complex and intricate, yet exact and highly reproducible, three-dimensional structures.

RNA is transcribed directly from the DNA, but proteins are made indirectly: first the DNA code is transcribed into an RNA message and then the RNA is shipped out to protein-making stations, called rough endoplasmic reticulum. Here, the RNA is translated into a protein sequence with each sequential group of three RNA beads directing the order of amino acid units that are strung on to the nascent protein chain. A single change in the DNA code could result in the protein chain being terminated prematurely or one amino acid being replaced by another. In turn, the altered sequence of amino acids in the protein chain may affect its function, changing body chemistry. Some of these changes give healthy variation; some are only important during medical intervention; some go beyond health and strain the system, but can be compensated by altering the amounts and activity levels of other proteins; and some changes fall outside these boundaries and cause disease or even death.

4.2 INTRODUCING ALLELES

The type of inheritance that pre-programmes obesity is often said to be the consequence of carrying a "fat gene" – a gene that some people have and some people don't and when you have it you become fat. However, in reality, most people, fat or thin, have exactly the same genes; it is actually the variety within genes that dictates body weight. The exact sequence of A, T, C and G beads in the genes inherited determine the exact recipe. The variety within genes is known as polymorphism, and the result of a polymorphism in a gene creates an allele. Alleles are different variations of a gene. Whether one confuses the term allele or gene may not seem so important, until we advance on to the next level of understanding, which is how the environment sways gene use to mimic the effect of a disease-causing allele. The low levels of leptin signalling that cause morbid obesity may be a consequence of inheritance, for example being dealt a variety of leptin gene – an allele – that reduces the efficiency of this signal, or the same effect can be mimicked by the environment, either via changes in gene use or directly via protein interactions. In both cases the impact on fat chemistry is similar, with the same outward manifestation, including both the deadly body weight and a overwhelming compulsion to eat. However, one cause is an inescapable consequence of genetic inheritance – the likely cause of Stone Age obesity – the other is a more fluid response to a changing environment, and is likely to be responsible for the majority of modern cases of the disease. In delineating these differences, the distinction between alleles and genes becomes quite important.

Most genes are paired – we have two copies, one inherited from each parent. And so we have the opportunity to inherit two different alleles of each gene. But within the entire human population there are usually many more than two varieties of each gene, giving all the subtle variations of different features that we see. In the case of the leptin gene, there are several alleles that give healthy variation, some alleles that signal for unhealthily low fat reserves, some alleles that predestine extra thick padding, and occasionally a combination of leptin alleles is inherited that causes obesity to a prodigious degree. The noticeable effects of inheriting particular alleles are known as the phenotype; the combination of alleles inherited is known as the genotype.

Understanding alleles is best illustrated with an example, and there is one example that most people know – the alleles that are major determinants of the compatibility of blood transfusions. There are over 70 different alleles of the *ABO* gene, each version with a slightly different DNA sequence, and most are just examples of healthy variety in the population. Some of these alleles have a noticeable effect on our bodies (perhaps only noticeable at a molecular level) and others do not, but some are the make-or-break determinants of compatibility in blood transfusions. The gene is named after the three most influential families of alleles that can be inherited at this site: A, B and O. Both the A and B groups of alleles make a mark on the surface of red blood cells, whereas the O alleles are noticeable by the absence of an effect. The *ABO* gene encodes the recipe for a protein called a glycosyltransferase enzyme. This protein transfers a glycosyl – a carbohydrate group – from a donor to a protein, modifying an amino acid building block by adding a carbohydrate tag. In the case of the *ABO* gene, the glycosyltransferase it encodes tethers carbohydrate units onto a protein called the H-antigen. The *A* allele recipe transfers a carbohydrate tag known as α-N-acetylgalactosamine onto the H-antigen, whereas the *B* allele product prefers tagging α-D-galactose in place. The *O* allele, however, does nothing of the sort: a single DNA base has been deleted from the gene sequence, which makes the recipe produce a non-functional protein and so no carbohydrate units are attached to the H-antigen. Once modified (or not, in the case of double *O* allele inheritance), the H-antigen ships itself to the surface of the cell and through the differences in the carbohydrate tag the body recognises its own red blood cells.

The *ABO* gene is paired, like many genes, and the body uses both copies simultaneously. Therefore, inheriting either two copies of an *A* allele or a single copy paired with an *O* allele give rise to the same blood group: such individuals can have different genotypes, either *AA* or *AO*, but the same phenotype, the blood group A. The same rationale applies to the B blood group. However, the O group is only apparent if two *O* alleles are inherited side by side. There is a fourth blood group that results from the combined effect of *A* and *B* alleles, the AB group. Therefore, your blood group is determined by the varieties of the *ABO* gene – the alleles – you have inherited and not whether you inherit the *ABO* gene itself.

There are many alleles that affect body weight. Obesity often runs in families: big parents produce big children. Several twin pairs separated at birth have been shown to grow up to have similar body weights.[4,5] Furthermore, twins tend to gain weight in a similar way – certainly more similar than random members of the population.[6] At first glance it appears as though these findings suggest that body weight is determined by the alleles we inherit, and so those who carry a genetic pre-determinant for obesity don't stand a chance of fighting fat. Such inevitability removes prejudice and blame from the disease, but it also defines a fate from which we cannot escape without medical intervention. And genetics is so amenable that even apparently random cases of obesity can be explained by inheritance.

When we consider inheritance, we usually look for similarities in features or traits rather than differences and back these observations up with phrases like "He has his dad's eyes". With this in mind, it may sound bizarre to consider genes as the cause of sporadic obesity like Lambert's when, in his case, only he in his family (and his town and his country) became so big. His condition may still have been caused by genes, but was sporadic because it was driven by genetic effects that can lie dormant for many generations. If it was, he is not the only example. More recent case studies of spontaneous, morbid obesity have been genetically characterised and shown to involve this type of recessive inheritance.

When inheritance is recessive, the effect of carrying an allele can be hidden by a dominant trait. Back to the ABO blood group, where the recessive trait is the O group. The presence of an *O* allele can be hidden behind the *A* or *B* alleles. Among those that influence body weight, alleles of the appetite-controlling leptin gene are particularly powerful. Some are able to trigger Lambert-style obesity – and even examples that make Lambert seem slim – but usually the most devastating examples are recessive and lie dormant for generations before popping up.

Even though they are elusive, recessive traits were identified before anyone knew what genes or alleles were. Back in the 19th century, the Austrian botanist Gregor Mendel sought to quantify the resemblance of children and their parents. True to trade, his subjects were plants, in particular pea plants, which allowed him to choose the parents and see the children grow to maturity within a year. Controlling parenting involved reducing the general promiscuity of the pea plant by carefully covering flowers to prevent any stray pollen from liaisons with the seed-making machinery. Removing the element of chance in pea parenting allowed the power of genes to be revealed. Mendel demonstrated and quantified the inheritance of features, or traits. For example, he discovered that among peas, short plants only ever made seed for short plants, and wrinkle shaped seeds only ever made plants that produced wrinkled seed. He postulated that the seed obtained immutable units of inheritance from the parent plants.

Mendel also discovered that not all of these inheritance units were the same; some were more powerful than others. It was he who named the more powerful ones "dominant", because they are always detectable; the others, called "recessive", lie hidden possibly for several generations. Take, for example, the seed shape. The allele that promotes a wrinkled shape is recessive and, therefore, masked by alleles for round shaped seed. A plant with a wrinkled seed must have two copies of the same allele, a genotype called homozygous, whereas a plant with round seeds has more options: either it can have two dominant round seed alleles or it may have just one paired with a hidden wrinkled version, making it hetero-zygous. Mendel went on to propose a mathematical model that showed that if both parents carried a single recessive allele and statistics were perfect, one in four of their children would inherit both copies of that allele and show the recessive trait. The others would display the dominant trait like their parents. Pairing allows the stronger, dominant allele to mask its partner's effect such that there is no visible trace of the recessive trait and it can lie dormant for many, many generations without trace until by chance it becomes paired with a similar recessive allele, and then its influence becomes clear. Such surprises are why a genetically predestined obese child can be born to trim parents.

4.3   SINGLE ALLELE OBESITY

Genes are the ultimate make-or-break system in body chemistry and fat chemistry alike: they are the basic unit of heredity, and without the gene that determines a particular feature, be it blue eyes or a trim physique, we just can't have it; at least not naturally.

If our inheritance predestines us to obesity then we will be big even if we would prefer to be slim. Inheritance can generate burning hunger that becomes an urgency to feed and to find food no matter what the dietary restrictions. Such cases of obesity demonstrate that inheritance is able to override the restrictions that are commonly believed to have kept our ancestors slim.

Geneticists have linked predestined obesity to the inheritance of specific alleles. Usually these alleles are non-functional versions of genes involved in growth, metabolism and appetite regulation. Beginning in 1996, a leading scientific journal in the field, *Obesity Research*, published an annual review of the status of the human obesity gene map. After a decade they pulled the plug; not because of a lack of data, as quite the opposite was true – the final version, in 2005, was enormous. The data encompassed over 100 pages in the journal. Compare this article to a more typical entry, where four or five pages is becoming the norm. The publication was divided into sections describing different types of factors that affect fat chemistry. With each gene there were examples of variations – polymorphisms – that altered body weight. Some of these genetic polymorphisms have more powerful effects than others. Some polymorphisms lead to a small amount of overeating, due to the inability of the body to judge its nutritional needs and detect a gradually growing girth, that leads to the obese status by midlife. These polymorphisms may be genetically encoded routes to the loss of compensatory appetite control. Other polymorphisms lead to an obsessive and overwhelming compulsion to eat, even when the body is already overfed. The most powerful consequences of this type of inheritance for body weight involve genes that usually protect against obesity. In many cases, the function of such genes is linked to regulation of appetite and metabolism. When varieties are inherited that are defunct, their power is lost and body weight climbs. And, at the risk of being repetitive, the leptin-encoding gene is a classic example of obesity protection that can fail.

The leptin gene was discovered in mice. Not any old mice, but a special type of mouse – a JAX mouse. JAX mice are the products of the Jackson Laboratory in Bar Harbour, Maine, USA. The mice in each group are bred so that, as far as possible, they are genetically uniform. In many cases their genotype is transparent from their phenotype and all their recessive genes are paired. The laboratory was founded as an independent, non-profit-making research institute in 1929 by Clarence Cook Little, a former president of the Universities of Maine and Michigan.[7] The site was named in honour of Roscoe B. Jackson, an engineer and the joint founder of the Hudson Motor Car company who had supported some of Little's research projects before his premature death from flu. The mission of the Jackson Laboratory was to develop healthcare to consider the unique genetic make-up of each individual.

Understanding the genetic make-up of individuals starts with understanding the genetic make-up of one individual – Mr or Ms Average. For people, there is no such thing as a perfect average; we all have quirks. However, animals can be bred to be much closer to perfect, whatever perfect is defined as. Perfect maybe an example of a specific breed – a Crufts champion – or it may be a perfect average, a true mongrel. Animals can be bred so that all progeny share selected characteristics, or traits, with their parents. For example, all Pekinese dogs have flat faces whereas all German Shepherds have longer muzzles. For some scientists investigating how genes make us what we are, even the variation between different individuals within a breed is too great; they strive towards genetically homogeneous models, because healthy variation within genes – alleles – can seriously interfere with observations of what different genes do and, because recessive genes can lie dormant for generations, their paired inheritance can introduce anomalies that skew scientific studies. Hence, research centres such as the Jackson Laboratories take breeding one step further to produce strains. A strain of animals is as genetically identical as a group of animals can be. Little worked with mice, and the mousey brothers and sisters he produced are genetically as similar as identical twins – but there are thousands of them.

4.4   OF MICE AND MEN

Mendel created pseudo-strains with his pea plants. He bred brothers and sisters together to produce pure strains with respect to one or another feature: height or seed shape, for example. Plants that bred pure with respect to the recessive trait are relatively easy to characterise because the genotype is transparent from the phenotype. In contrast, pure-breeding with respect to the dominant allele is trickier because the recessive allele can hide, and remain hidden for many generations, only becoming apparent when it is coupled to its equivalent. The quick test to discover hidden alleles for shortness is to cross-breed a tall pea plant with the transparent genotype of a short plant. Seeds germinating and growing into short plants are evidence enough for a hidden short gene. Using such controls, Mendel managed to achieve his goal.

Clarence Little continued where Mendel left off. He produced pure-breeding strains with respect to all traits: living libraries of defined genotypes – at least as defined as Nature allows. There are now many strains of organisms, from bugs to furrier beasts, but Little made one of the first: a strain of mice named DBA (Dilute, Brown, and non-Agouti). He bred littermates together for several generations to purge the family of hidden recessive traits and to produce many mice as identical as monozygotic twins.

In addition to a higher level of genetic homogeneity, strains differ from breeds in that breeds are selected for beauty (in the eye of the breeder at least); in contrast, strains can be far from pretty. At the Jackson Laboratories, humanity defies natural selection by allowing mice with various abnormalities, some of which would not survive outside the laboratory, to breed. Breeding occurs in a controlled manner with the aim of developing mouse models of human diseases. Despite being far from the fluffy pets mouse fans usually desire, the demand for the Jackson Laboratory's mouse strains is higher than the demand for fancy mice. Marketing JAX strains, as they became known, has financially supported the research laboratories – even through the 1929 stockmarket crash – right through to modern day.

The high demand for strains centres on the fact that mice and humans have many diseases in common, and at a biochemical level they become more rather than less similar. Thus, sick mice can help scientists to understand disease progression in humans and also to develop drugs to combat these complaints. JAX mice are used to understand and develop treatments for obesity. Although the answer to the obesity question remains elusive, these mice have vastly advanced other areas of medicine. Currently the various critters in the collection include models used to understand, cure or treat birth defects, Down's syndrome, osteoporosis, AIDS, anaemia, autoimmune disease, tissue transplant rejection, athero-sclerosis, diabetes, gallstones, hypertension, blindness, cerebellar disorders, deafness, epilepsy, glaucoma, macular degeneration, neurodegenerative diseases and a range of cancers such as bone, cervical, liver, mammary and ovarian cancers, leukaemia and lymphoma. For each disease there is a highly developed strain of genetically defined mice that serves as a model. More than 4000 inbred mouse strains are estimated to be housed at the Jackson site – the accumulation of many years of work. However, this accumulation has not always been in an upwardly direction because fires have twice swept through the animal houses. The leptin story began when the Jackson Laboratories were recovering from the first such incident – recovering both from the shock that such a thing could happen and also attempting to recover speci-mens for their living mouse library so as not to lose years of research into human disease completely. The recovery was inad-vertently facilitated by their sales programme: exporting mice to other laboratories acted as insurance, such that when the onsite stocks were incinerated, specimens from the original stocks, along with newcomers, were shipped back from all corners of the globe. Among these immigrants was a morbidly obese mouse that trig-gered a whole new wave of research.

Usually, selecting an animal model of a human disease begins by screening animals for the right symptoms and then confirming that their disease matches that of the humans. The leptin story occurred in reverse: it began when an unusually fat mouse was found amongst the more typical mouse-sized specimens. The mouse was big enough to be pregnant, but with no sound of pattering paws, and close inspection revealing the mouse to be male, he was given the uninspiring name of obese, "ob" for short. He fed frantically, pawing at the chow dispenser long after his meal time was over. He consumed three times more than the average mouse. The atypical mouse behaviour coupled to the atypical mouse physique attracted researchers' attention, and work began to find out exactly why this fellow behaved the way he did.

The ob mouse became big by eating the same type of chow as other mice and in the same environment, suggesting that the cause of his blubber was genetic. The next step towards identifying the culprit was to develop a pure-breeding strain with this affliction: a family of supersized mice. Simply said, but enormously challenging because the female ob mice were sterile. Ob progeny could only be produced using elements of *in vitro* fertilisation (IVF) technology and surrogate mothers. Despite these challenges, the lure of understanding obesity fuelled the research effort for decades until the so named *ob* gene was found. In a poignant twist (possibly with a view towards commercial exploitation), rather than calling the protein produced from the gene "obese", in line with the name of the mouse and the gene, it was named leptin – derived from the Greek word *leptos*, meaning thin. The naming is more apt, because the *ob* gene product does not make mice fat: it makes them thin. Only when this gene becomes defunct does mouse size verge on massive.

At the simplest level, leptin can be considered as a "fat-stat": just like a thermostat regulates heating by closing down the heater when the desired temperature is reached, leptin is produced in proportion to fat deposits and closes down feeding when appropriate reserves are in place. Most leptin is produced in the white fat tissue under the skin and around organs, but smaller amounts are produced in other areas of the body, including fat tissue, glands, bones and the gut. The ob mice do not produce leptin at all. Their leptin-producing alleles are defunct varieties; polymorphisms have abolished their function. They frantically feed because, despite the fat deposits that are killing them, their bodies are stuck in starvation mode. They pull out all the stops to find food, overcoming the normal need for rest and play.

The leptin story was launched into the public realm very visually at the end of 1994 with pictures on the front cover of the prestigious general science journal *Nature* showing a healthy mouse dwarfed by a huge leptin-deficient cousin. At this time, the details of the role of leptin in human appetite regulation were not fully understood, but extrapolating the fact that a lack of leptin leads to obesity in mice, one could conclude that a little more of this hormone in humans could reverse body weight gain and lead to a more desirable trim physique. Thus, the story stirred interest in the pharmaceutical industry – the race was on to elucidate the role of leptin in human body weight control.

Jeffrey Friedman of the Rockefeller University, New York, USA, was one of the many scientists who became interested in the ob mouse. He set out to find out exactly why these mice became so obese. The smart money was on a polymorphism in a regulatory hormone gene, but exactly which gene remained to be seen. Using a technique known as positional cloning, which is a method that searches for a gene by its location on the chromosomes rather than its function, Friedman tracked down the *ob* gene and then went on to make the leptin protein.[8] Once the gene had been identified, it was relatively straightforward to find spelling mistakes that converted the normal *ob* gene into a recipe that failed to produce functional leptin. The double-sized mouse was attributed to a single change in the genetic code, termed a point mutation: a single letter spelling mistake in the *ob* gene stopped it being used. Just like their trim cousins, the huge, leptin-deficient mice still have an *ob* gene, but their allele does not produce functional leptin so these mice eat themselves into an early grave. In humans, several genetic changes that completely block leptin function have been mapped. One is a result of a point mutation that causes message reading to be aborted prematurely, resulting in a shorter and non-functional leptin protein, and another change in the DNA sequence causes the leptin message not to be read at all.[9] Alleles have been identified that lead to the production of non-functional leptin and obesity,[10] all the way through to polymorphisms that just cause variety in leptins rather than disease.[11–13]

The defunct leptin gene is recessive. Such genes could be present in any member of the population, lurking there silently until paired with another such gene to make its host obese. This is what happened in the case of the ob mouse. The first ob mouse found at the

Jackson Laboratories was not the first mouse to carry the new allele; he was the first mouse to carry it in a double dose. Despite his family having been subject to extensive breeding programmes to remove recessive genes, at some time in his ancestry Nature allowed the spontaneous change in a leptin gene to give rise to the new allele, which was then passed on to the next generation. Quite when this change happened is difficult to say because the hidden character of recessive genes means simply that: it was hidden. Thus, the mutation in the *ob* gene that super-sized the mouse may even have occurred while his ancestors were housed at the Jackson Laboratories, but only after shipment and return was the gene paired and the mouse huge.

As with mice, within the human population there have been examples of spontaneous mutations that create a recessive allele that causes disease only when paired; these diseases include obesity. Since molecular medicine was born, these diseases can be accurately diagnosed by the exact change in the DNA code and, like mouse leptin genes, the human equivalents include recessive alleles that cause morbid obesity only when paired. Recall the case of the leptin-deficient cousins. These children were amongst the first patients to receive such molecular detective investigations for obesity. The cousins were born to a consanguineous Pakistani family. After only a few years of life, they travelled to Cambridge, UK, for help with their weight.[14] The journey may seem extreme, until you learn about their obesity problem: the eight-year-old weighed in at 89 kg (14 stone) and the two-year-old nearly 30 kg (4.5 stone). The hidden nature of the disease meant that by chance it affected both cousins, but not their siblings; the obesity-promoting alleles were recessive and so were hidden for generations. Stephen O'Rahilly, a professor of metabolic medicine at the University of Cambridge, found that the children's blood contained very little leptin. He and his research group went on to map the problem at the level of the genetic code. They found that there was a single nucleotide deletion in the cousins' leptin gene that affected the whole code thereafter, changing the nature of the next 14 amino acid residues in the protein, and the final 21 were missing altogether.

Just as machines often only work when all component parts are functioning, proteins are much the same. The shape of the protein positions the working parts correctly so that it can function. In the case of a machine, altering the shape or component parts slightly may still allow it to function, but the efficiency may change – perhaps it will be less efficient or it may even work better than before. Alter the machine significantly and it will not perform the original function. The same is true for proteins: small alterations in protein shape may affect efficiency, yet significant changes abolish their specific functions. The repercussions of such changes depend on both the gene affected and the severity of the effect. Small changes in less significant proteins give rise to healthy variation within the population; large changes in significant proteins cause disease. The importance of protein function means that genes make us what we are largely via proteins: if a part of the body is not protein, then it is made by a protein because some are chemical machines called enzymes that, along with a handful of important RNA machines, perform all the tasks needed for life. Proteins are involved in management, manufacturing and communication.

When the leptin signaller is sufficiently damaged to abolish its function and its appetite-suppressing effect, patients are always hungry. The two cousins ate non-stop from birth, panicking when food was restricted. The fault in the leptin signalling pathway resulted in more than half their body weight being fat. Just like the leptin-deficient mice, the children were morbidly obese and their lives were in immediate danger because of their weight. Once all other avenues had been tried, the doctors decided to treat the cousins with daily doses of leptin, and guess what? The leptin boost suppressed their appetite and these children lost weight.[15]

When there is a fault in the leptin signalling pathway, fat reserves are not registered and hunger is not quenched. Giving leptin-deficient individuals – mice, men or kids – injections of the missing hormone simulates a functional fat gauge and causes body weight to drop. But synthetic leptin is not an anti-obesity drug for the wider population: the treatment only works when the body does not produce enough leptin of its own, and such conditions are rare. Therefore, the rare cases of leptin injections quenching appetite are the exception rather than the rule.

Although more general trials of synthetic leptin demonstrated that defunct alleles of the *ob* gene are not central to the obesity epidemic, this does not rule out the possibility that other recessive, obesity-promoting alleles have been lurking, hidden within the population waiting to pop up in the next generation (nor does it exclude the role of leptin in the modern obesity epidemic). Just as powerful as defunct *ob* genes are some alleles of the *db* gene. As with leptin deficiency, this disease was first identified in mice with an abnormally large body weight. Unlike the ob mice, these critters were also diabetic. In wider studies, it has been shown that diabetics often have low levels of leptin production, promoting weight gain,[16] yet blood tests showed that the db mice had plenty of leptin in their blood. This test ruled out the *ob* gene as a candidate for the disease, yet leptin signalling was still involved. It is now known that the *db* gene encodes for the receptor switch through which leptin mediates its effect. On binding to the receptor switch, leptin induces a shape change, which transmits the appetite-quenching signal. Mice with completely defective leptin receptors, such as the db family, are as fat as the ob mice, because despite producing ample leptin, the hormonal signal cannot be relayed.[17] There are many polymorphisms of the *db* gene, some of which have a greater effect on signalling than others. Some give rise to healthy variation; others cause disease. For example, one of these changes involves deletion of a single G nucleotide in the *db* gene – just one letter is omitted from the gene – and the result is a 10-fold reduction in leptin receptor efficiency. Mice carrying a double dose of this allele die before their 10th month.[18] Still, the genetic polymorphism remains in the mouse population because it is recessive and this fatal flaw can move through the generations, hidden.

There are several well characterised genes that are recessive and that cause obesity, such as the alleles of the corticotrophin-releasing hormone receptor (CRHR). The genes that carry the recipe for the CRHR protein reside on two human chromosomes: 7 and 17. At both sites, polymorphisms can promote obesity, but usually only when inherited in a double dose do they trigger increased release of adrenocorticotropic hormone (ACTH), which in turn raises cortisol levels, causing body weight gain and eventually Cushing's disease. In the case of the obesity-promoting alleles of the melanin-concentrating hormone receptor 1 (MCH_1) and melanocortin receptors (MC), the effect is mediated via the altered function of the signaller's switch rather than the signaller itself. Just as leptin works through a molecular switch or receptor encoded by the *db* gene, other molecular signallers have their own switches through which they pass on their message. When either the switch or the signaller malfunctions, the same signs and symptoms can be apparent. For example, when either leptin or its switch is knocked out, the individual becomes morbidly obese. These and other powerful obesity-promoting alleles are likely to have been the cause of the majority of obesity cases throughout the first and second stages of nutritional history: the cases of the disease that defied calorie restriction, exercise regimes, steam rooms and the salt baths most commonly prescribed to those that could afford the physicians' fees. These pre-determinants were rare; hence the low incidence of obesity. They are just as rare today. And they always will be because of natural selection.

4.5 NATURAL SELECTION

Although the term natural selection is familiar to most, it is shrouded in misunderstanding. Natural selection is a simple theory that describes how certain features become more or less common within a group of individuals. The concept is based on the assumption that the environment offers limited resources for living things: limited types of food, water and shelter. Given these limits, not all the individuals born will survive to adulthood; some will fall by the wayside. Those that do survive are the best of the bunch with respect to reproducing in the current environment. If the environment changes they may lose their advantage. They are not necessarily the fittest or the healthiest (although these attributes may help their cause), but they are the ones that can produce the most genetic heirs.

Natural selection states that individuals that are best adapted for reproducing, perhaps because they have a feature that gives them a little advantage or do not have a feature that could be a disadvantage, have more babies with this same profile. The more offspring they have, the greater the percentage of apt individuals in the next generation and gradually the special advantage will become the norm. This theory can be extrapolated further to suggest that over a sufficiently long period of time, and with pretty brutal culling of individuals, natural selection filters the diversity within a group to create a new species that is so good at doing what it does it can survive where newcomers fail. As communities become isolated in different locales, each will be subject to a different selection pressure and each community will develop characteristic traits. The diverse species on Earth and the exquisite features many possess that make reproduction in their particular habitats that little bit more successful is proposed to be a result of this natural selection process. Of course, across the surface of planet Earth there are dramatically different environments. Natural selection only provides pressure for a species to adapt for survival in its current niche and if the species moves or the environment changes, the adaptations may become maladaptations and the species may become extinct.

Humans are not exempt from the natural selection theory. Only recently have industrialised nations become privileged enough to expect their children to survive to adulthood; in other parts of the world, childhood mortality and morbidity are, unacceptably, more in line with the Stone Age statistics. And while there is competition and culling there is selection. Each generation in each locale becomes streamlined, becoming more genetically alike and marginally better suited to the environment in which they live. With respect to body weight, the selection pressure pushes in several directions. On one hand, natural selection offers protection against the spread of genes that promote the disease state of obesity: morbid obesity reduces fertility and life expectancy, preventing reproduction and the continuation of the disease into the next generation. The more severe the disease, the rarer it will be. A dominant disease that completely blocks reproduction will self-extinguish, whether this is a result of premature death or infertility, and obesity attacks from both sides. Even if an allele is recessive, it is still subject to the same pressure: in their silent state there is no force to change their frequency and when they become paired the line is ended abruptly. So natural selection provides pressure for the elimination of alleles for morbid obesity, whether dominant or recessive. On the other hand, other alleles that promote unhealthy body weight gain, perhaps in middle age, have more opportunity to feature in the next generation and when a little energy conservation, greed or laziness helps reproduction, natural selection can help these traits become more common.

4.6 MALADAPTATION

In 1962, James Neel, a geneticist at the University of Michigan Medical School, proposed that some disease traits we see within a population are abundant because they gave an advantage for survival at some point in history.[19] Neel's work was primarily on diabetes. He rationalised the abundance of genes that predispose to diabetes by suggesting that in one environment, perhaps one lost in history, the varieties of genes that promoted the development of diabetes gave a reproductive advantage – a case where the selection pressure is not equal and favours the mild disease state. His logic was simple and attractive and was soon expanded outside diabetes. There are many examples of features that help survival in one environment yet cause disease in another, of which sickle cell anaemia is possibly one of the easiest to understand.

Sickle cell anaemia is most commonly the result of a point mutation in the genetic code that informs the production of the oxygen-carrying blood protein haemoglobin: the *HBB* gene on chromosome 11. Haemoglobin, densely packed into aptly named red blood cells, is what makes blood red. In the lungs, this protein picks up oxygen from inhaled air and carries it around the body to where it is needed to release energy from food. The oxygen is converted to carbon dioxide, which is transported back to the lungs dissolved in the blood to be released in exhaled air. Many polymorphisms in the haemoglobin gene affect the efficiency of its oxygen-carrying capacity, most of which cause healthy variation, but some can cause disease. In the case of sickle cell anaemia, the most common form results from a change in the *HBB* gene that directs the replacement of a large positively charged glutamate amino acid residue at the sixth position in the protein chain with a smaller, hydrophobic amino acid called valine. This substitution may seem small, but it has a significant effect on haemoglobin, altering the structure of the protein and reducing its ability to carry oxygen.

The effect of the substitution on the oxygen-carrying capacity of the protein does not cause sickle cell anaemia in itself, but the altered protein can distort the shape of red blood cells under certain conditions. Normally, red blood cells are disc-like with slight depressions in the centre, a shape that optimises the flow of oxygen in and out of these cells and aids circulation through the blood vessels. However, cells carrying sickle cell haemoglobin change under low-oxygen conditions to a more crescent-like shape. During the lifetime of a sickle red blood cell, it will travel about 100 miles around the body in blood vessels, cycling through rounds of sickling and desickling as the oxygen concentration fluctuates, and eventually becoming irreversibly sickle-shaped. In this form the cells have a habit of blocking the tiny capillaries that distribute blood to individual cells in the body, starving them of oxygen. The result of oxygen starvation is cell damage or death. The result of losing a few cells depends on which cells are lost: we probably would not notice the loss of a few muscle or skin cells, but important brain cells or hormone-producing cells in the glands may cause problems that are more serious. Still, sickling can have its advantages.

Not all genes follow Mendelian inheritance, as the dominant–recessive rules are now called. For example, the *ABO* gene only partially follows Mendelian inheritance. The A and B traits are both dominant over O and inheritance considering either A or B with respect to O follows Mendelian maths, but A and B are also equivalent to each other and as soon as the second dominant gene is factored into the equation, Mendelian inheritance rules no longer apply. Yet Mendel found enough examples in pea plants to propose a generalisation. He probably studied many inherited characteristics before he found the handful of examples needed to underpin his theory of inheritance, which at the very least suggests he was a patient and determined researcher. However, such extensive studies would have revealed to him that his inferences, although correct, were not exclusive. He was undeterred by the many exceptions and thus, the man was either incredibly stubborn, incredibly talented or divine intervention was guiding him (he was, after all, a monk). It seems that Mendel knew what he set out to prove before he even proved it – not the best attitude in science, but in his case it was successful.

Like the *A* and *B* alleles, the sickle cell haemoglobin allele and the more usual variant of the gene do not exhibit Mendelian inheritance. Both alleles affect the phenotype of the individual. A single copy of a sickle cell allele coupled to a normal haemoglobin allele can offer protection against malaria, without debilitating disease. A small amount of sickle cell haemoglobin mixed through the usual variety causes sickling and either rupture or leakage of the red blood cell when the oxygen concentration is low, which can happen for many reasons, including malaria infection. The leakage is lethal to the parasite.

The distribution of the sickle *HBB* allele across the world maps onto the distribution of malaria: carrying one sickle cell anaemia haemoglobin allele can cause problems during exercise, especially when the oxygen levels in the air are low, but the protection this polymorphism offers against malaria is thought to have promoted its abundance in parts of the world where this parasite is common. Applying the same rationale to genes that promote obesity, one could hypothesise that carrying alleles that predispose to this disease in one environment may aid survival in others. For example, alleles that predispose to obesity in modern, industrialised cities may give the edge on reproduction where the food supply is unreliable.

Just about every country in every corner of the world can cite a time of recent famine, even those countries who believe they are now immune to such a problem. This pressure is not new, but the tail-end of a long survival struggle. Emerging from the last Ice Age put an extra strain on our ancestors; strain that readily culled the weak and frail individuals. Such an environment is fertile ground for natural selection and adaptation, accelerating the spread of traits within a population at the expense of similar, but less efficient, characteristics. When food is in short supply, one of the desirable characteristics is the ability to stockpile fat such that the detrimental health effects of future famines are buffered. Still, even in such an environment, the powerful promoters of obesity could not spread because they cause too severe a disease. In contrast, the less powerful genetic polymorphisms – those that only strengthen the tendency towards obesity, perhaps by loosening compensatory appetite regulation or even abolishing it when the reproductive years draw to a close – have more potential to spread, and spread widely, when many generations are exposed to the same selection pressure – hunger. The thrifty genotype theory depends on the existence of such alleles and long-term food shortages, and the human story has both. The widespread nature of shortages suggests a global distribution of the thrifty genotype, but in some countries the shortages have been harder for longer, and attempts have been made to correlate the incidence of obesity among various populations with the hunger their ancestors experienced. Sometimes the data fit.

The Pacific Ocean is the world's largest water body and extends from Singapore to Panama. Within the waters are almost

800 habitable islands including the Polynesian, Melanesian and Micronesian islands. These archipelagos have approaching the most unstable natural food supply in the world and are often densely populated. They were originally inhabited by Micronesian and Polynesian peoples, with Europeans arriving through the 18th and 19th centuries. Although they brought some nutritional stability via imports, the newcomers also brought a different kind of turbulence through wars, occupation and bombing. Extrapolating Neel's theories to encompass the incidence of obesity would suggest that such longstanding selection pressure towards fuel economy would promote the accumulation of the varieties of genes that promote fat stockpiling. Just as the thrifty genotype theory predicts, now regular imports are being received, the people are suffering.

Each island has its own character and cultural diversity, but the peoples migrate between islands and so share similar genes. These genes have shouldered much of the blame for their obesity: many of the Pacific Islands have the highest incidences of obesity in the world. One, Nauru, is the world's smallest republic and, globally, it has the biggest obesity problem. It tops the fat polls with over 90% of the adults weighing in obese,[20] not just overweight but actually obese. Here being big is normal, despite the negative health implications. Among the Nauruan people there are certain alleles of the leptin-encoding gene that are common,[21] although how these variations affect body weight has yet to be established.

Kosrae is another Pacific Island. It boasts a land mass of 42 square miles and a slightly lower pudge statistic: here every other adult is obese. Still, you don't need to be right at the top of the fat polls to attract the attention of people who may be able to help,[22] and Kosrae's friendly scientist is Jeffery Friedman, the man whose detective work found the *ob* gene. He has applied his super-sleuth skills to identify the cause of obesity among the Kosraeans, and in one study involving over 2000 adults a direct link was found between staying trim and the degree of European ancestry.[23] With the thrifty genotype in mind it is possible to speculate that the unpredictable Micronesian climate has selected for the thrifty genotype, making the people more susceptible to obesity now shop shelves are packed. The Europeans, however, arrived more recently and so have experienced less of this pressure and therefore such ancestry offers protection against obesity. Under the thrifty

genotype hypothesis, the Kosraeans and the Nauruans are mala-
dapted for modern living, whereas the Europeans are more
adapted.

In other parts of the world, other alleles have been identified that
could promote or protect against unhealthy body weight gain.
Polymorphisms in the perilipin gene are thrifty genotype candi-
dates. Perilipin is a lipid droplet surface protein present in the
specialised fat-storing adipocytes and steroid producing cells.
There are at least five different polymorphisms in this gene that are
linked to body weight, and their individual abundance within a
population depends on ethnicity. For example, certain weight-gain
promoting variations are more common in Indians than the
Chinese. According to the thrifty genotype theory, such differences
in allelic abundance suggest that the Indians have had a less stable
nutritional history. The distribution of these alleles is notable in
Singapore. On this collection of about 60 islands neighbouring
Malaysia and Indonesia, most residents are of Chinese origin. Just
over 77% of Singaporeans are Chinese, about 14% Malays and
half that number are Indians. Obesity is more common in Malays
and Indians compared with the Chinese. Probing the differences
between these groups on a molecular basis shows that the Malays
and Indians share polymorphisms in the perilipin gene that are less
common in the Chinese population; carrying such polymorphisms
more than doubles their chance of becoming obese.[24] The obesity
statistics are paralleled on the mainland: in China, the incidence of
obesity is low, averaging less than one in twenty. In contrast, India
has an epidemic. One could tentatively conclude that the Indians
have spent longer adapting to shortages and are now more mala-
dapted than the Chinese. In addition to causing obesity when
food supplies are unlimited, these same alleles may protect against
future famines. The protective advantage has yet to be established,
and the incidence of these alleles has yet to be mapped onto the
distribution of past famines, but the polymorphisms in the perilipin
gene may still be contributing to the apparent maladaptation of the
human race.

One of the most concerning candidate genes for the thrifty
genotype is one with a very long name: nicotinamide adenine
dinucleotide-dependent deacetylase silent mating type information
regulation 2 homologue, or sirtuin for short. This protein has
multiple functions, and therefore it is difficult to pinpoint which

one is the key to body weight control; all may be involved. At the level of signalling, a lack of sirtuin in the pro-opiomelanocortin (POMC) neurons abolishes obesity resistance. In fact, individuals become hypersensitive to diet-induced obesity: they lose compensatory appetite regulation. In addition to eating more they are unable to buffer fat reserves through energy wastage. Such observations suggest that sirtuin mediates its anti-obesity effect via POMC signalling. Moreover, when sirtuin is eliminated, energy expenditure drops such that little overindulgences cannot be buffered and the excess is stored as fat. The effects extend even further: in mice that inherit defective sirtuin genes, leptin fails to engage signalling in POMC neurons properly. These mutant mice also show altered fat deposition. In particular, they are less able to control visceral fat stores:[25] they are prone to belly flab, promoting the health problems associated with obesity at a lower than expected body weight. These factors include the increased production of ACTH (see Stone Age Obesity: Of Mice and Men) and cortisol, further promoting visceral fat accumulation and declining health. Therefore, the alleles that give rise to low sirtuin production bias the energy balance towards body weight gain, change fat deposition to induce obesity at a lower than expected BMI, and abolish compensatory appetite regulation – all the features of the modern obesity epidemic can be explained. And alleles that raise sirtuin function lead to obesity resistance.

However, there is a worrying undertone: recall that sirtuin is also linked to ageing. Via sirtuin, the complications attributed to body fat, for example cancer, neurological degeneration and diabetes, may not be a consequence of obesity, but part of the same cause. And if changes in sirtuin function are causing the obesity epidemic, fat is a symptom of a much more worrying disease state.

4.7 A ONE-WAY STREET

A quick glance at the past suggests that the obesity epidemic is an innate and inevitable consequence of hardship, and this view has been popularised. The idea is based on the assumption that the human gene pool spent many years adapting to natural living – years accumulating alleles that promote body weight gain – and has not changed since. In fact, it is estimated that much of the population was malnourished with respect to energy right up until the beginning of the 20th century; for almost the whole of human history individuals who could stockpile fat may have had a survival advantage such that natural selection has ensured that most of us have some elements of this tendency today. Yet needs have changed: modern technology cocoons us in cosy homes with ample food and advanced medical care, at least in industrialised nations. Energy is now abundant and it is up to the individual to regulate their food consumption. Many people are failing in this challenge. We could blame the thrifty genotype as a maladaptation: our genetic design is a mismatch with the environment we have created. In addition to leading to obesity, this mismatch has been blamed for many other diseases, including cancer, stroke, heart disease and diabetes, to name but a few of the so-called diseases of civilisation.

It is worth reiterating that natural selection only provides pressure on a species to adapt to an environment by selecting those individuals best able to reproduce; it does not increase fitness as such. Therefore, if the environment changes, features that were advantageous can quickly become a downfall. The thrifty genotype could be the downfall of the human race. Although they may have aided survival in the hunter–gatherer communities where the obesity-promoting effects were suppressed by food restrictions, when the restrictions were lifted these same polymorphisms could cause an epidemic of disease. And so it seems that natural selection has shaped the obesity epidemic; it seems that natural selection is determining human destiny.

It is also worth remembering that natural selection is a constrained and clumsy process: constrained by the rate of change within our genes and clumsy because it depends on the completely random process of spontaneous mutation. The speed and randomness means that fate directed by natural selection alone can see a species die out before it adapts. This is an important point: even if an adaptation could be made to improve survival, it cannot be made unless, by chance, the right polymorphism spontaneously occurs.

In the modern world, the consequences of carrying the obesity-promoting polymorphisms that could be considered as part of the thrifty genotype do not significantly affect reproduction, and so there is no natural selection pressure to reverse the accumulation of obesity-promoting alleles. True, obesity reduces lifespan, but with social support systems and medical technology, on average it only trims the last nine years, and, in general, even though people are leaving it later to have children, the premature end comes when the childbearing period is over. True, the obese are more likely to suffer from fertility problems, but on a background of everyone having fewer children and modern medical technology to iron out some of the blips, the obese produce about as many offspring as the trim. Taken together, these observations mean that even if natural selection made us thrifty enough to cause an obesity epidemic when food restrictions were lifted, it has no power to reverse the obesity trend now shop shelves are full.

4.8 GLOBAL MALADAPTATION

On the Pacific Islands, European ancestry has been linked to the ability to resist body weight gain.[23] If we assume for a moment that ancestry only affects genes and not culture, these data seem to suggest that the unpredictable Micronesian climate has selected for the thrifty genotype, whereas the more stable European climate has allowed a more frivolous calorie usage without any detriment to survival. If lifestyle and calorie consumption are about the same among all people on the island, irrespective of ancestry, now shop shelves are packed with imported supplies, the thriftier genotype of the native islanders renders them more susceptible to obesity compared with the people whose ancestors lived with a more stable food supply for longer – generally those countries supplying the imports.

The thrifty genotype nicely explains the obesity problem among the populations of some countries, but does the rationale hold if we go global? Of all the peoples in the world, the Sentinelese are likely to have the least stable food supply, war aside. They live off a particularly fickle example of the fat of the land because their island is located in a tsunami zone. When one of the most recent hit, it was with such a force that observers believed that the Sentinelese people may have met their demise, until their helicopters flew through a cloud of arrows; there may have been fatalities, but not enough to distract from the fight against intruders. Such uncertainty at mealtimes suggests that their dose of thrifty genes is likely to be the greatest in the world. Observations, although limited, have yet to spot a glutton. Yet one could argue that the food supply is still too limited to allow the fat consequences of carrying a thrifty genotype to become evident, and if these islanders ever decide to welcome unlimited food supplies onto their sandy shores, one could predict that obesity will become the norm.

The thrifty genotype theory suggests that the more hunger our ancestors experienced, the more likely we are to become obese today. Our ancestors experienced hunger for many different reasons and unstable weather patterns that would have wiped out crops are a hot favourite for promoting a thriftier genotype among the Pacific Islanders. With respect to weather patterns, they experience more extremes than the Maltese, who, in turn, are more likely to be battered by storms rolling across the Med than the Latvians, who are simply caressed by the tranquil Baltic Sea. The storm report correlates with the incidence of obesity; still, this is not the complete picture. Storms are not the only cause of lean years. Hunger is also caused by relying too heavily on one crop type that is then obliterated by the spread of a disease; examples are said to include the Potato Famine in 19th-century Ireland and, more recently, in 2007 black stem rust obliterated wheat crops in many countries including Uganda, Kenya and Ethiopia. Famine is also a consequence of war and many people are still experiencing hunger due to conflict. Cold can augment the consequences of hunger too, because of the greater amounts of energy consumed in keeping warm. Colder climates require more food energy to be transformed into heat and, therefore, more nutrition is needed to achieve the same body frame as in a warmer, friendlier climate. When these factors combine, it is difficult to see how the genotypes of the various European nations could have been pressured by hunger to such different extents that they can now explain the demography of obesity.

The Maltese are part of the European ancestry that is said to protect the Pacific Islanders from obesity; it just does not seem to work on the home continent. OK, so the incidence of obesity in Malta is less than the incidence in the Pacific Islands, but they have an unacceptably high level of disease compared with the other countries in Europe, with whom they share similar weather patterns, food availability and genomes. Archaeological evidence suggests that humanity first arrived in Malta from Sicily around 5200 BC. These hunter–gatherers eventually died out, or at any rate disappeared, and were replaced by the Phoenicians from Tyre around 1000 BC until the late 8th century BC, when a Greek colony was founded on the island. Every 200 to 600 years power changed hands again: Carthage, Rome, the Vandals, Rome again, the Fatimids, Sicily, Spain, Rome again, the Ottomans, and the French sequentially commanded the island. By the 1800s, Malta gave up going it alone and voluntarily became part of the British Empire. With each invasion there was genetic drift; natives may have been slaughtered and newcomers arrived. Still, the genetic make-up of the modern-day Maltese suggests that although half of the Y-chromosomes have Phoenician origins,[26] such that these males can trace their ancestors back to the inhabitants three millennia ago, the majority of the Maltese genes have Southern Italian influences, with a little input from the Eastern Mediterranean.[27] From these foundations, natural selection has had plenty of opportunity to increase the incidence of the thrifty genotype, because right up until the beginning of the 20th century the Maltese were hungry; hungry enough to riot in 1919.

To add a further question mark over the theory that the thrifty genotype promotes obesity, consider the Balts. The Balts have had their fair share of occupation too, although they started out from a different gene pool. Skeletons of the first settlers have been dated from 6000 BC. Around 4 AD they began to lose land to Slav invasions, but managed to remain a culturally unique people. Latvia and Lithuania were occupied by Poland, Sweden and Russia, before gaining independence in 1918, only to lose it 22 years later as a result of Soviet occupation that lasted for the next five decades. During the various occupations hard times were known, yet the energy balance in these countries suggests that rather than selecting a thrifty genotype, polymorphisms were selected that offer obesity resistance.

The Balts were subject to an additional selection pressure: the cold. One would expect that the further north Europeans settled, the stronger the pressure towards the thrifty genotype because of the climate. Although farming may have been more stable in northern Europe than, for example, in the Pacific Islands or perhaps even Malta, the closer to the poles the colder it becomes, providing pressure towards thrifty energy use in the same way as a lack of food availability. Lithuania is in the north-east of Europe, next door to Latvia and the Baltic Sea. In contrast, Malta lies right at the south of Europe, somewhere close to the tip of Italy's toe.

Accordingly, the evolutionary pressure exerted by the colder climate to promote fuel economy would affect the Latvians and Lithuanians more than the Maltese. Yet the Balts appear to have resistance. With the same type of pressure, how would natural selection drive the gene pool of one population one way and another in the opposite direction?

A genetic basis for obesity removes individual responsibility for the disease, but creates an inescapable fat-trap; inescapable unless one is prepared to suffer the Stone Age hardship of hunger while surrounded by a sea of nutrition. Given the human race's general dislike of hunger, it seems as though obesity is here to stay. Yet, it is worth remembering that every allele identified as part of the thrifty genotype is a variation in a gene that functions in fine-tuning body weight to survival – a gene which in its most common form buffers body weight in the healthy zone. For example, amongst its many roles, the leptin protein acts as a fat-stat to protect against unhealthy body weight gain until its function is lost. Similarly, the alleles of the CRHR, ACTH and perilipin genes that have been correlated with unhealthy body weight gain are rare examples of variants of genes that offer protection against these same symptoms. In each case, more people have the alleles that offer body weight protection than promote stockpiling.

The average calorie intake throughout most of human history was less than what is considered ideal today. Yet battles against excessive body weight began two millennia ago. Such data suggest that humans tend to stockpile fat. On the other hand, over the past few hundred years, industrialised living in the UK has offered many people a calorie excess. Perhaps there was not enough energy to cause an obesity epidemic, but there was enough for more people to cross a BMI zone or two if they desired. However, the vast majority of the population chose to be slim. It was not an easy choice; many people struggled with their weight – certainly enough to allow Banting's success to give him celebrity status – but there were no obesity epidemics. The desire to be slim remains; the success statistics have changed. The power behind modern obesity is on the same par as the genetic diseases that once were the most common cause of obesity, but is a loss of compensatory appetite regulation and a biased energy balance part of our innate programming or is it a consequence of the environment we have created in following this same programming?

The thrifty geneotype theory is dependent on an environmental change to render adaptations as maladaptations. This change is the critical factor in the obesity epidemic. Consider the Japanese – champions at staying trim and obtaining longevity, at least until this new millennium. Japan has known frequent storms, both natural and political, that have seen people go hungry. It has also had an ample food supply for several decades and yet evidence of obesity is only just beginning, coincidentally with a change in diet and lifestyle, including food imports and franchises from the West.

Obesity among the Japanese and the Balts suggests that, although the story of natural selection-based obesity is convincing and can be illustrated with seemingly apt examples, the environmental trigger is more important than the alleles. As a final thought, consider the Chinese: although there is a low overall incidence of the disease in China, the fat people are clustered in the cities. It is safe to assume that, irrespective of their postal address, the Chinese have about the same distribution of genetic variation, but city-living is causing a health problem: over one-fifth of the folk living there are obese.

4.9 OBESITY RESISTANCE

The thrifty genotype theory describes a scenario in which the future human physique looks set to grow, but scientists are increasingly unveiling another group of alleles that are less easy to spot. The obesity-promoting alleles only became noticeable in recent decades when the environment changed to promote unhealthy body weight gain; similarly, these more recent discoveries remained indistinguishable in the population until unhealthy body fat became the norm and their presence became glaringly obvious by the resistance they offer.

Some people do not easily stockpile fat. The *rikishi* actively try to gain weight: some are more successful than others; few are as successful as the populations on the Pacific Islands. In Micronesia obesity has hit hard, affecting the vast majority of adults. Yet if the predisposition to unhealthy body fat is attributed to inheritance, it is equally valid to suggest that those that do not become obese have varieties of genes that offer obesity resistance. Similarly, over in Europe the variation in obesity suggests as much resistance as thrifty energy use. Even among slim nations there is evidence for obesity resistance: the Latvians and Lithuanians have similar ancestry and exercise levels, and, as expected, the incidence of obesity is similar in both countries, but the average Lithuanian has available a huge calorie excess,[28] demonstrating obesity resistance.

Some rats and mice, like people, have protection against obesity: when faced with meals of high-fat chow, some rodents are able to adjust their metabolism and food intake to prevent unhealthy weight gain. There is a strain of mice known as Lean, because they are indeed lean.[29] Exactly why they are lean and stay lean is not clear. These mice produce more in-house cholesterol-synthesising equipment and have more of the protective high density lipoprotein version of this fat circulating in their blood; they also produce more bile acids from this steroid. Bile acids have many functions, one of which is in the digestion of fat; the Lean mice are not protected against obesity because of an inability to digest or process fats. Quite the reverse is true, because the raised production of bile acids compared with the average mouse makes the process more efficient. This is an interesting consideration for anti-obesity drug development, because many current strategies focus on blocking fat uptake. Rather, Lean mice have altered muscle energy metabolism, which promotes increased physical activity and thermogenesis – heat generation. Their protection is deep-rooted in their fat chemistry, and the altered energy metabolism seems to be inherited. The polymorphisms that cause mice to be energetic have yet to be identified specifically, but seem to be in the regulatory elements of DNA that control the way our bodies use their genes. The mice do not have novel appetite regulatory or metabolism controlling alleles, but the way they use the more common varieties of genes differs enough to offer obesity resistance. Using techniques such as positional cloning – the same technique as Jeffery Friedman used to track down the leptin gene – the inheritance has been linked to a particular stretch of DNA, but the exact gene and mechanism involved have yet to be pinpointed.

Some strains of mice can fight fat by increasing the number of leptin receptors as their body weight rises, maximising the amount of signal detected[30] and protecting against unhealthy weight gain. Yet the Lean mice work their magic in a different way, showing that there is not just one but many mechanisms of obesity resistance. Sirtuin alleles are another example: low levels of sirtuin are linked to obesity; high levels cause obesity resistance. The profile of their effect matches what we know of the current obesity epidemic: individuals lose compensatory appetite regulation and cannot buffer reserves through energy wastage. They lose the Lean mouse's protection.

4.10   ACHIEVING BALANCE

The incidence of obesity around the world suggests that energy balance is achieved with disproportionate loads of food and exercise. This observation is grounds for the thrifty genotype theory, but can be explained with or without the accumulation of energy-economy alleles. In fact, food supply fluctuations have been buffered throughout human history without epidemics of obesity or anorexia, suggesting that fat chemistry can achieve energy balance in a surprising number of circumstances. Recall Britain through the ages. Back in the days of King George III, which is when Lambert's legacy began, food was not restricted enough to keep the entire population slim: cooks became podgy, wet-nurses were well padded and gentleman sometimes stout; yet obesity was rare. There are many documented cases of individuals who became heavy and then lost weight. People took control when they were too heavy, using the principle of energy balance. Most were successful and those that failed usually did so because of underlying disease rather than greed or laziness. If the thrifty genotype and ample energy were all that was necessary for obesity, one would imagine that as food supplies rose there would have been a growing number of competitors for the title of "fattest man alive", but Lambert won by a considerable margin. Obesity was attainable but, generally, people were just plump and usually because of middle-age spread. Through to the 1950s the story was much the same: there was enough food for people to overeat, but average calorie consumption was less than the current recommended daily intake. With the higher activity levels than today's generation, the energy balance was tipped in favour of weight loss, but our ancestors did not waste away to nothingness. Part of the deficit was balanced by their smaller frame, but it is also likely that their fat chemistry adapted to the shortcomings with thrifty energy use: not a thrifty genotype but a transient adaptation in fat chemistry that biased the energy balance so that it was tipped in favour of weight gain, perhaps by adjusting the metabolic rate. Close to the other end of the distribution, consider the Latvians today. They exercise through their work, but leisure time is as sedentary as possible, yet they still manage to balance an energy excess. They achieve energy balance through changes in fat chemistry: the extravagant mode. It is likely that energy use has oscillated between thrifty and extravagant modes in this and other populations over generations – perhaps within generations. But right now they are able to burn a bagel of calorie excess.

A little overindulgence can be balanced by tipping the energy balance in favour of weight loss, for example by mounting more blocks to represent energy use due to increased body temperature. Adaptations in fat chemistry are not implemented under any circumstances, only to protect the body against unhealthily decreasing or increasing fat reserves. This was a logical survival strategy for Stone Age living, allowing us to remain healthy despite a fluctuating food supply, but one that flies in the face of the popular belief that we are designed to stockpile whenever possible. It demonstrates how our bodies are protected against unhealthy weight gain. It also expands the daily calorie allowance from a single figure to a range. Our bodies can survive fit and trim with a little less than we need, as demonstrated by the 1950s Britons. We can also burn slight excesses in energy, as in the case of modern-day Latvians. In both situations, the difference in calorie consumption is similar, but there has to be a limit to how far our bodies can sway their chemistry. The Lithuanians seem to have a super-high level of protection and are able to bias the energy balance sufficiently to balance an energy excess equivalent to a portion of fries. People in other countries, such as Mexico and the Pacific Islands, seem much more vulnerable to weight gain. The same bias in energy balance can be found within countries: some people can adapt their body chemistry to achieve energy balance sufficiently to prevent obesity, and some do not. And when fat chemistry fails, obesity is the result.

4.11 WHERE FAT CHEMISTRY FAILS

Malta's turbulent history was not least because the islands are in a prime location for expanding trade routes through the Mediterranean Sea. In the mid-1990s, there was significant nutritional transition on this island as a result of trade liberalisation with the creation of the World Trade Organization (WTO). As a consequence, the relative cost and availability of foods changed. Meat fell in price, vegetables became available all year round and there was a flood in the supply of spam and sugar (both in the granulated form and hidden in drinks and condensed milk). Faced with a convergence in the availability and cost of different food groups the islanders were able to overindulge their genetically programmed taste preferences, favouring easy to eat processed foods, animal products, oils and fats, sweet foods and refined grains.[31]

Malta is not alone in this recent dietary change. In the countries worst hit by obesity there has been a similar transition. Jump over the ocean to Nauru. Unlike Malta, this island is too small and not particularly well placed to benefit European shipping. Geography explains why for many centuries this and other Pacific Islands were left alone by all but a few whalers and pirates, until the last century. Geography also explains why Nauru became an official stop-over for immigrants heading for Australia. Limited natural resources make the population dependent on food imports, and the foods imported are surprisingly similar to those that reach Maltese shores: sweet, fatty and salty foods perfectly matched to our genetically programmed taste preferences. Similarly, Kosrae, closer to Hawaii, became a strategic and much needed military post for the US forces. Distinct from many military operations that subsist on basic food rations, when the USA deploys troops they are supported by the fast-food flavours of home. Like the Maltese and the Naurauns, Kosraeans were introduced to fast food – and to white rice, sugar and spam. And, like the Maltese, somehow the natural foods, in this case fish, breadfruits, bananas and coconuts, became significantly less popular.

Similar changes in diet and obesity are documented for the Chileans and the Pima Indians of Arizona. Before the 1950s, these populations subsisted on a varied diet of meat, fish, fruits and vegetables. However, aid in the form of high-fat and highly salted foods has been accompanied by a steep rise in the number of cases of obesity.[32–34] In the last decades, many countries have seen dramatic dietary change. China is a prime example, where Western food imports offer an international feel to the cities, but have yet to permeate into the countryside. Latvia and Lithuania have undergone rapid redevelopment in the same period, yet floods of sugar and spam have yet to reach the shores of the Baltic Sea. Japan and Indonesia, where turbulent food supplies could be considered more on a par to those in the Pacific Islands, are only just beginning to be affected by the Western diet and obesity is only just becoming a problem.

On the Caribbean island of Bermuda, which is estimated to have amongst the top five fattest populations, officials are being bold in apportioning blame: two-for-one strategies used to promote the sales of sugar-sweetened breakfast cereals are cited as one cause. Then again, oversized breakfasts are just part of the problem. Take a seat on Church Street, in the centre of the capital, Hamilton, on a weekday lunch and see the people flock out of their offices to buy deep-fried food. Although many of the familiar fast-food giants have yet to site a franchise, the island has its own just as fatty equivalents. Fast food has become the norm, usually washed down with a conscience-clearing fruit punch, deceptively laden with sugar.

Rather than historical hardship, the incidence of obesity correlates more closely to current dietary change. It seems as though there is something about modern energy-dense nosh that seems to cause us to gain weight. Calories are a prime contender, but cannot take all of the blame. Something in this dietary transition seems to have caused fat chemistry to fail, pretty much independently from genetic inheritance.

REFERENCES

1. M. J. Hudson, M. Aoyama, T. Kawashima and T. Gunji, *Anthropol. Sci.*, 2008, **116**, 87–92.
2. H. von Staden, ed. C. J. Palerne, Publications de l'Université de Saint-Etienne, Cedex, 2003, vol. 9, p. 869.
3. H. Pearson, *Nature*, 2006, **441**, 398–401.
4. A. J. Stunkard, J. R. Harris, N. L. Pedersen and G. E. McClearn, *New Engl. J. Med.*, 1990, **323**, 1069.
5. A. J. Stunkard, T. I. A. Sorensen, C. Hanis, T. W. Teasdale, R. Chakraborty, W. J. Schull and F. Schulsinger, *New Engl. J. Med.*, 1986, **315**, 130.
6. C. Bouchard, A. Tremblay, J. P. Despres, A. Nadeau, P. J. Lupien, G. Theriault, J. Dussault, S. Moorjani, S. Pinault and G. Fournier, *New Engl. J. Med.*, 1990, **322**, 1477–1482.

7. A. M. Ingalls, M. M. Dickie and G. D. Snell, *J. Hered.*, 1950, **41**, 317–318.
8. Y. Y. Zhang, R. Proenca, M. Maffei, M. Barone, L. Leopold and J. M. Friedman, *Nature*, 1994, **372**, 425–432.
9. J. Hager, K. Clement, S. Francke, J. Raison, N. Lahlou, N. Rich, V. Pelloux, A. Basdevant, B. Guy-Grand, M. North and P. Froguel, *Int. J. Obes.*, 1998, **22**, 200–205.
10. M. K. Karvonen, U. Pesonen, P. Heinonen, M. Laakso, A. Rissanen, H. Naukkarinen, R. Valve, M. I. J. Uusitupa and M. Koulu, *J.Clin. Endocrinol. Metab.*, 1998, **83**, 3239–3242.
11. L. Oksanen, K. Kainulainen, M. Heiman, P. Mustajoki, R. Kauppinen-Makelin and K. Kontula, *Int. J. Obes.*, 1997, **21**, 489–494.
12. K. Silver, J. Walston, W. K. Chung, F. Yao, V. V. Parikh, R. Andersen, L. J. Cheskin, D. Elahi, D. Muller, R. L. Leibel and A. R. Shuldiner, *Diabetes*, 1997, **46**, 963.
13. N. Matsuoka, Y. Ogawa, K. Hosoda, J. Matsuda, H. Masuzaki, T. Miyawaki, N. Azuma, K. Natsui, H. Nishimura, Y. Yoshimasa, S. Nishi, D. B. Thompson and K. Nakao, *Diabetol*, 1997, **40**, 1204–1210.
14. C. T. Montague, I. S. Farooqi, J. P. Whitehead, M. A. Soos, H. Rau, N. J. Wareham, C. P. Sewter, J. E. Digby, S. N. Mohammed, J. A. Hurst, C. H. Cheetham, A. R. Earley, A. H. Barnett, J. B. Prins and S. Orahilly, *Nature*, 1997, **387**, 903–908.
15. W. T. Gibson, I. S. Farooqi, M. Moreau, A. M. DePaoli, E. Lawrence, S. O'Rahilly and R. A. Trussell, *J. Clin. Endocrinol. Metabol.*, 2004, **89**, 4821–4826.
16. S. Gulen and S. Dincer, *Mol. Cell. Biochem.*, 2007, **302**, 59–65.
17. M. Maffei, J. Halaas, E. Ravussin, R. E. Pratley, G. H. Lee, Y. Zhang, H. Fei, S. Kim, R. Lallone, S. Ranganathan, P. A. Kern and J. M. Friedman, *Nature Med.*, 1995, **1**, 1155–1161.
18. J. A. Brown, S. C. Chua, S. M. Liu, M. T. Andrews and J. G. Vandenbergh, *Am. J. Physiol.*, 2000, **278**, R320–R330.
19. J. V. Neel, *Am. J. Hum. Genet.*, 1962, **14**, 353–362.
20. *Obesity in the Pacific: too big to ignore*, Secretariat of the Pacific Community, 2002.
21. A. M. de Silva, K. R. Walder, T. J. Aitman, T. Gotoda, A. P. Goldstone, A. M. Hodge, M. P. de Courten, P. Z. Zimmet and G. R. Collier, *Int. J. Obes.*, 1999, **23**, 816–822.
22. D. E. Duncan, *Technol. Rev.*, 2005, **108**, 52–59.

23. E. Ruppel-Shell, *The Hungry Gene*, Atlantic Books, London, 2002.
24. L. Qi, S. Y. Tai, C. E. Tan, H. Q. Shen, S. K. Chew, A. S. Greenberg, D. Corella and J. M. Ordovas, *J. Mol. Med.*, 2005, **83**, 448–456.
25. G. Ramadori, T. Fujikawa, M. Fukuda, J. Anderson, D. A. Morgan, R. Mostoslavsky, R. C. Stuart, M. Perello, C. R. Vianna, E. A. Nillni, K. Rahmouni and R. Coppari, *Cell Met.*, 2010, **12**, 78–87.
26. C. Franklin-Barbajosa, in *The National Geographic*, National Geographic Society, Washington, DC, 2004.
27. C. Capelli, N. Redhead, V. Romano, F. Cali, G. Lefranc, V. Delague, A. Megarbane, A. E. Felice, V. L. Pascali, P. I. Neophytou, Z. Poulli, A. Novelletto, P. Malaspina, L. Terrenato, A. Berebbi, M. Fellous, M. G. Thomas and D. B. Goldstein, *Ann. Hum. Genet.*, 2006, **70**, 207–225.
28. I. Janssen, P. T. Katzmarzyk, W. F. Boyce, C. Vereecken, C. Mulvihill, C. Roberts, C. Currie and W. Pickett, *Obes. Rev.*, 2005, **6**, 123–132.
29. M. Simoncic, T. Rezen, P. Juvan, D. Rozman, G. Fazarinc, C. Fievet, B. Staels and S. Horvat, *BMC Genom.*, 2011, **12**, 96–108.
30. S. Lin, L. H. Storlien and X. F. Huang, *Brain Res.*, 2000, **875**, 89–95.
31. B. Popkin, C. Albala, S. Benjelloun, L. Bourne, G. Cannon, D. Coitinho, C. Doak, O. M. Galal, H. Ghassemi, G. Harrison, V. Kosulwat, M. J. Lee, T. Maletnlema, V. Matsudo, C. Monteiro, M. I. Noor, K. S. Reddy, J. Rivera, A. Rodriguez-Ojea, R. Uauy, H. H. Vorster and F. Y. Zhai, *Public Health Nutr.*, 2002, **5**, 279–280.
32. D. Shmulewitz, S. C. Heath, M. L. Blundell, Z. H. Han, R. Sharma, J. Salit, S. B. Auerbach, S. Signorini, J. L. Breslow, M. Stoffel and J. M. Friedman, *Proc. Nat. Acad. Sci. USA*, 2006, **103**, 3502–3509.
33. P. Zimmet, P. Taft, A. Guinea, W. Guthrie and K. Thoma, *Diabetol.*, 1977, **13**, 111–115.
34. E. Ravussin, P. H. Bennett, M. E. Valencia, L. O. Schulz and J. Esparza, *Diabetes Care*, 1994, **17**, 1067–1074.

CHAPTER 5

The Image of Fat

Reading a title like "The Image of Fat", one could be forgiven for thinking that this chapter is all about the persona of excess body weight. No, this chapter is all about the image of fat itself – molecular fat. Exploring the image down at an atomic level and understanding the significance of the arrangement of atoms for the properties of the molecule helps one to appreciate the merits and menaces of this substance.

As far as its merits go, fat is remarkable stuff and has an exclusive place in society. The *Encyclopaedia Britannica* gives a definition of fats that enables these materials to be distinguished from others: "any substance of plant or animal origin that is non-volatile, insoluble in water, and oily or greasy to the touch", and with such sensory distinctions it is easy to identify fats, until they become incorporated into food products, because too much grease is unpleasant enough to stimulate the development of cooking methods – natural or otherwise – to mask the slimy feel of fats.

The properties of fat have given it a largely unique place in food and industry for many years and, in several cases, fat binds these domains together. One of the first industries to be built around fat was candle making. Starting in monasteries, candles were made by coiling the fat known as beeswax around a wick, and the by-product from the manufacture of candles was the foodstuff honey, sold to the public to raise additional funds. The principal ingredient in the detergent and soap industry is also fat, traditionally

Fat Chemistry: The Science behind Obesity
Claire S. Allardyce
© Claire S. Allardyce 2012
Published by the Royal Society of Chemistry, www.rsc.org

tallow, a by-product of meat processing. Hundreds of years on, many products have fat as an ingredient and many more depend on the other functions of fat. Consider grease. Grease is the name for fat used as a lubricant (and a hair product); it is also the verb that describes smearing it around. This practice dates back over 3000 years, when fat and lime were combined to make axle grease. Most moving parts in machines are greased and, because most manufacturing processes depend on machines, fat could be considered as a fundamental ingredient in manufacturing in general. Similarly, oil is another name and action linked to lubrication, but this alias is used more widely to include waterproofing and the names of liquid fats sometimes used as fuels, as materials used in the manufacturer of paints and varnishes, and as ingredients in our food.

Yet most of the aliases of fat do not refer to its essential roles in industry and modern living, rather they are names given to body fat. Lipid is the specific term scientists use to describe fats in living things, and adipose cells, organised into tissues bearing the same name, are where the largest lipid stores reside. But words like blubber, bulk, flab and the dreaded cellulite are more familiar. We even have names for specific areas of body fat, such as thunder-thighs, bingo-wings, jelly-belly and love-handles; all are synonymous with body fat. Most of these names are mocking, ridiculing part of ourselves because – just as every circus once had a fat lady (usually married to a skinny man) – we still see fat as a spectacle. The number of aliases reflects the level of human fascination with body fat. It so interests the world that Lambert is still considered a medical marvel, and new understandings of his biggest symptom are often headline news. Despite the impending energy crisis, the discovery of a new pocket of oil underground may cause a localised media buzz, while the discovery of a potential weight loss product creates a worldwide hullabaloo.

The preoccupation with body fat is perhaps why the general understanding of this substance has become chronically biased. Most people are aware of the universal energy storage role of this material, but its other functions are much less widely known. Within the group of fats named lipids, certain members have unique properties that give them irreplaceable functions in the body. In fact, fat has the most fundamental function of all: it defines the boundary between life and the rest of the world. And it can form these boundaries because the simple versions of fat molecules are insoluble in the most abundant substance on Earth – water.

Most of the world is covered in water. In that water various things dissolve – the two become one – even when one material starts out as a solid. Take table salt, a crucial part of modern cooking, as an example. This mineral is known by chemists as sodium chloride. In its pure form, it is a solid at room temperature, which is handy for sprinkling on chips, but when a spoonful of salt is added to a glass of water it dissolves without a visible trace; it is completely soluble. Hidden salt gives seawater its characteristic flavour and, in addition to making up the sea, salt-water makes up most of our bodies too. In this salty environment our bodies perform all the functions necessary for life. Back in our kitchen cupboards we can find the carbohydrate most people call sugar and scientists call sucrose. A spoonful of this sweet stuff is soluble in a glass of water too, disappearing without a trace. Sucrose can be cleaved into fructose and glucose; fruit sugar and the body's preferred fuel, respectively. Glucose and fructose dissolve in water, allowing easy transport around the body in the watery part of blood called the plasma. Other substances are efficiently carried in blood plasma too: the glucose-regulating hormone signaller insulin is an example. This messenger is neither salt nor carbohydrate, rather it is protein. These differences demonstrate that many different types of molecule are compatible with water – hydrophilic – but fats are not.

Most simple fats, such as plant oils and lard, do not dissolve in water. They are completely hydrophobic – water-shy. They are water-shy because water and fat are fundamentally different: whereas water is polar, simple fat is not. Magnets are polar, having a north and south pole at opposite ends. Similarly, a water molecule has poles, but not the magnetic type; rather they are formed because of an uneven distribution of charge across the molecule. A water molecule is made up of two hydrogen atoms joined to a single oxygen atom. The latter is eight times bigger than the former and so the molecule looks a bit like Mickey Mouse's head (or Hello Kitty's, if you prefer), where the face represents an oxygen atom and the ears are the two hydrogen atoms. Oxygen is bigger than hydrogen because it has more positively charged protons (and also neutral neutrons) in its nucleus. In addition to the nucleus packed with protons and neutrons, atoms are wrapped in an electron cloud, and when they join to form molecules parts of the clouds merge. The big oxygen nucleus has a strong pull on the electrons that are orbiting around the molecule, drawing them towards itself so that they linger longer over this part of the molecule. The uneven charge density means that the hydrogen atoms are slightly positively charged and the oxygen carries a slight negative charge. The uneven charge density also allows a type of interaction, called hydrogen bonding, where the opposite charges attract. Hydrogen bonds stretch from one water molecule's Mickey's ear to another's chin. The attraction is strong enough to hold water together and give it unique properties, including its liquidity at room temperature when most other substances made of similar sized particles float away as gas.

In addition to holding water together as a liquid, the polarity determines what can dissolve in it and what cannot. At the simplest level, any molecules that can be accommodated in the hydrogen-bond network, even if the pattern has to change slightly, can dissolve. Salts that are composed of small charged particles can slip into the network relatively easily and their charge allows them to interact with the patches of charge on the water molecules. The particle units of signallers such as insulin are huge, yet they can still dissolve in water if they carry enough patches of charge to maintain enough of the network to keep water happy. Molecules that have no charge cause disruptions. Even when such molecules are mixed and shaken with water to allow a blending, the troublesome network-disrupters are gradually evicted as the pattern reforms. Simple fats do not have enough charge to be accepted into water's gang.

Mix fats with water and each type of molecule clusters with its own kind to give some very interesting effects. Emulsions, which include milk and some paints, are opaque because tiny fat droplets are suspended in a watery medium without truly mixing. In some cases, these tiny fatty droplets fuse together repeatedly to form ever increasing globs and eventually all become one big blob that floats to the top. Mixing the fat and water phases back together recreates the emulsion at least temporarily; with the help of chemical technology and mechanical homogenisation, the emulsion can be maintained for the long term. Mixing is helped if the water or fat is modified. Vinegar is a watery substance containing acetic acid. Acetic acid belongs to the fat family, but is too small to be completely immiscible with water. Instead, the intermediate nature of acetic acid helps oil to become suspended in the vinegar, for example when we make a salad dressing. Still, vinegar alone is not powerful enough to suspend fat in water for long periods of time, which is why we tend to shake homemade dressings before we pour.

In addition to making salad dressings tricky to make, but tasty, the tendency of fats to shy away from water can be inconvenient. It is because fats and water generally do not mix that using water alone to clean a dish that held a dressed salad or, worse still, a roasted joint of beef, is a struggle. The fat can be transferred to a cloth or a basin, but only a little is removed because fats shy away from water. Enter detergents and soaps. Like acetic acid, these cleaning products bridge the gap between water and fats. They combine some molecular properties of fats with a polar group to persuade the water and fat to mix so that the latter can overcome its water-shyness and get carried away.

Soaps have been part of our lives for over 5000 years,[1] often made by modifying the fats and oils themselves. Early accounts suggest that the modification process involved mixing grease with ashes to produce a basic cleaning agent – basic in the simple sense and basic in the chemical sense. In chemistry there are acids, like vinegar, and bases, like bicarbonate of soda. Bases are also called alkalis. Acids are sour; bases are bitter. Both can burn. The burning sensation of acids and bases is like a sting. In fact, wasp stings are basic and bee stings are acidic. When you mix an acid and a base together they react to produce water and a salt. Remember all that information and you can use vinegar to neutralise the wasp sting by turning the basic venom to salt and water. On a bee sting, bicarbonate of soda – a base – does the same trick. The degree of burning depends on the strength of the acid or base: stings are not that strong, just injected into a painful place. Similarly, vinegar and bicarbonate of soda may tingle, but hardly burn. In contrast, battery acid and bleach are examples of a strong acid and base, respectively, and can cause serious damage to our bodies, partly because they can disrupt the function of body fat, turning it into soap.

The importance of soap cannot be understated; candles may have been outdated by the introduction of electricity, but soap is still used daily by most of the population. It even saves lives.

In 1847, Hungarian-born Ignaz Semmelweis noted that childbed fever, a major cause of illness and death for women in the maternity ward, could be prevented if the physicians washed their hands before examining their patients. He has since been accredited with revolutionising the understanding of sterilisation in medicine.[2] Given that most doctors in that era spent their spare time between delivering babies brushing up on anatomy in the morgue, a bit of carbolic soap revolutionised patient care.

Soap is so highly valued that the story of its discovery has become part of folklore. The Italian version of the "How soap got its name" story is set at the foot of the sacred Mount Sapo. According to the legend, rain would wash the mountain clean, including both the animal fat and ashes from the sacrificial fires. That mixture would then enter the River Tiber. On discovering that clothes washed cleaner downstream, people flocked to the mountain's foot to do their laundry. When the cleaning product was finally identified, it was named after the mountain. Although the legend is likely to be largely fiction, not least because no record of a place with this name appears in the history of Rome, the recipe for making soap is true. The ashes and fat mixture remained relatively unchanged for centuries, independent of the continent. For example, American colonists stuck to the animal fat recipe, using tallow mixed with potash solution collected from their winter fires, whereas some Europeans branched out to use olive oil as an alternative fat. Today the process is similar, but chemistry has added a little of its know-how to make the process harsher, more effective and far more complicated. Customers are no longer satisfied with simple cleaning, but demand perfumes, moisturisers and other special ingredients to buffer the less desirable properties of crude soap, including its corrosive basicity.

5.1 MOLECULAR DETAILS

The molecular arrangement of fats gives them their characteristic properties. Both the nature of the atoms and how they are linked together is important in making these substances non-volatile, immiscible in water and greasy to the touch. On one level, these properties are a consequence of the fact that fats are mostly made of two types of atom: hydrogen and carbon. When just these two types of atom are strung together in chains they go by the name of hydrocarbons. In addition, fats contain a dash of a third type of atom, oxygen. All the oxygen – a total of two atoms, regardless of the total number in the molecule – is located at one end of the chain to form what is called a carboxylic acid group, the very same that is responsible for the sourness of vinegar and its ability to mix with water, due to the uneven charge distribution across this assembly of atoms. The remainder of the molecule is formed from a chain of carbon and hydrogen atoms intuitively called a hydrocarbon or, less intuitively, an acyl tail. Together the carboxylic acid and the acyl tail form a fatty acid, one of the fundamental units of fats. As the acyl chain grows in length, the contribution of the carboxylic acid group to its taste and hydrophilicity is gradually diluted and the hydrophobic properties of the tail take over. Yet despite its flavour being masked, the carboxylic acid group has critical importance, including chemical characteristics that allow facile modifications to tailor the properties of the fat to specific functions.

The principal types of atom in fats are the same as in another shunned nutrient, carbohydrate, but the ratios of the atoms and their organisation within the molecules are different, which is why these molecules have different properties. In the case of carbohydrates, these molecules are so named because the ratio of atoms is as though the carbon is hydrated – as though water has been added. For each carbon atom there is one oxygen atom and two hydrogens. The arrangement of the atoms is not water-like, but similar enough to maintain their affinity with water because of localised polarity.

The usual arrangement of atoms in carbohydrates involves building chains of six carbon atoms and adorning them with six oxygen atoms and twelve hydrogens. Each carbon atom has the potential to make four links – it has four electrons to share. Recall that electrons are given, received or shared when atoms form the chemical bonds that hold molecules together (see Why the Fuss about Obesity? Clearing the Name of Chemistry). Therefore, if two electrons are used to make the chain, two remain to link other atoms and form branches along it. When a hydrogen atom is linked, because it only has one electron it finishes the branch. In contrast, oxygen atoms have two electrons so when linked to a carbon atom they still have one to share with another atom, for example a hydrogen cap to form what chemists call a hydroxyl group. The arrangements of hydrogen and hydroxyl groups along the chain determine the properties of the carbohydrates: both glucose and fructose contain the same number of carbon, hydrogen and oxygen atoms, but their arrangement within these molecules results in quite different properties in our bodies.

In fats, the carbon atoms are usually arranged in a chain, just like carbohydrate carbons, but the number of atoms is not limited to six, extending down to just two and up into the thirties. Along this chain the free electrons on the carbon atoms gather only hydrogen atoms. The lower oxygen content of fats compared to carbohydrates has two principal effects. First, it makes fats energy dense. Energy in food is loosely proportional to the amount of oxygen that can be added. Compared with fats, carbohydrates intrinsically have more oxygen added per carbon, such that there is less oxygen-adding potential and less trapped chemical energy to be released by the body. Second, carbohydrates are also lower in stored energy per gram than fats because they carry a lot of water lodged in their structure – less oxygen makes the fats more hydrophobic. When a hydrogen and oxygen link to form a hydroxyl group, the electrons become localised around the larger atom, giving an uneven distribution of charge. These hydroxyl groups can liaise with water molecules and form part of the network. In contrast, fats only have an uneven distribution of charge over the carboxylic acid head. This part of the molecule can form hydrogen bonds with water, but the hydrocarbon tail has a uniform charge distribution and so it cannot. For small members of this family, such as acetic acid, the charge on the head compensates for the tail's aversion to water, allowing the two types of molecule to mix. But as the tail becomes longer, its properties dominate and just as the water network shuns fat, fats reciprocate.

Hydrophobicity can be an advantage in a water-based body, for example allowing fats to make boundaries and form fixed fuel supplies that do not float away. Yet it can also be a disease-causing disadvantage because such molecules clump together to form blobs that have the habit of blocking body parts. The body can tailor the hydrophobicity of fats for specific functions using chemical modifications. These modifications usually take place at the oxygen-carrying head of the fatty acid molecule. One common example is increasing the hydrophobicity that aids storage by tethering several fatty acids together across a special linker called glycerol. Glycerol is a small molecule with three carbons; each acts as a docking point where fatty acids can be attached using the carboxylic head as molecular glue – the result is what is termed an acylglyceride (AG). The linker is the consistent unit in all AGs and is relatively small. Its three points need not all be filled, such that one, two or three fatty acids can dock to form a monoAG, diAG or triAG (TAG), respectively. Linking fatty acids via a glycerol bridge facilitates their transport and collection in the body. It also increases the viscosity of the material because the combined weight of three carbon tails means that in a pure form it tends to be a thick liquid or even a waxy solid, rather than a free-flowing fluid.

Despite fatty acids having the same overall pattern – a hydrocarbon chain with a sticky carboxylic acid tethered to one end – their properties vary dramatically. The size of the chain is one factor that dictates their properties, and it extends from just two carbon atoms right up to thirty-six: from vinegar to tar. Chains mostly form regular, repeating zigzags of carbon atoms. If there are fewer than six carbons in the row, the fatty acids are termed "short". The smallest of these molecules are truly carboxylic acids, rather than fatty acids, because their tails are too short to give the characteristic hydrophobicity of fats. The very shortest is acetic acid, the acid in vinegar, and the longest in this group is called caproic acid. Caproic acid is sometimes classed as a "medium chain fatty acid", a group that encompasses the members of this molecular family with carbon tails of six to twelve atoms. It has a six-carbon chain tail and, like those fatty acids with eight and ten carbon atoms, derives its name from the Latin work *caper*, meaning goat. The naming alludes to their goat-like smell and the fact that they represent the most abundant fatty acid in the milk from this animal. These fatty acids are not exclusive to goat products; they are found in coconut milk, for example, but unless they form the bulk of the fat, the strong smell is masked by other molecules in the mix.

Most foods are composed of an assortment of fatty acids such that the blend gives the overall characteristics. However, some foods are rich in a particular type and then its individual characteristics can be detected above the background of the others. Goat's milk is one example; palm oil is another. Around half the fatty acids in this oil are the long chain sixteen-carbon variety called palmitic acid and so its properties contribute strongly to the properties of coconut fat. Still, palmitic acid is found in other fats too. In fact, it is one of the most common fatty acids found in nature.

Palmitic acid is a major component of cocoa butter, along with stearic acid and oleic acid. These three types of fatty acid are tethered onto glycerol linkers – one at each port – and the properties of this TAG dominate the properties of cocoa butter: it is a solid at room temperature, but, because its melting point is close to body temperature, it can easily liquefy on a warm day and also melt in the mouth.

Very long chain fats form more waxy or tar-like substances. Beeswax is indeed a wax because it contains a high proportion of the twenty-six carbon fatty acid cerotic acid. Lignoceric acid has a twenty-four carbon atom chain and is found in wood tar, a substance traditionally used in medicine and to waterproof boats and sails and still used as flavouring and scent in sweets, saunas and cigarettes. These long chains are so big and hydrophobic that they can form waxes without the need of a glycerol bridge.

5.2 THE SIZE OF FAT

The length of the carbon chain determines the usefulness of fatty acids to the body: the short to medium length fatty acids are readily absorbed and digested, providing vital energy to people whose guts are not able to take up as much nutrition as they need from their food[3] and a readily available energy source to people who need to gain weight, for example because of a wasting disease.[4] Yet medium chain fatty acids are also claimed to be protective against unhealthy body weight;[5] they are easily absorbed by the liver and less easy to store than the long-chain alternatives, which is thought to promote their direct use as a fuel rather than their accumulation. Several weight loss regimes are based on this idea, but given that the function of fat is not all about metabolism and energy, such diets do not always consider all aspects of health. Fats are important constituents of the body, from the waterproofing waxes on our skin through to the neurological functions of the brain.

Whereas short and medium chain fatty acids are readily absorbed by the gut, longer members of the family need to be modified before they can cross the intestinal lining, and the modification forms part of the digestive process. The long fatty acids have more than twelve carbon atoms in their chains and the very long fatty acids more than twenty-two. Their hydrophobicity means that in the stomach they do not mix with the watery digestive juices. The body produces bile acids to persuade the fats and juice to mix. Like soap, bile acids are made from modified fats. In this case, the starting material is neither tallow nor olive oil but cholesterol. In the liver, hydrophilic groups are attached to the hydrophobic starting material to equip it for its role in facilitating digestion. The unmodified molecule is mainly hydrophobic, with a single hydroxyl group tethered on one side. This property allows it to bridge the gap between water and fat and it does so in some foods. Butter is mostly made of the same type of fat as lard, yet the texture of the two materials differs: one is spreadable on a warm day and the other is harder and more likely to crumble. One of the differences between the two food products is the water content: about one-sixth of the composition of butter is water; lard has none. The water is held in place partly by the cholesterol in this fatty food. Cholesterol enables the tiny water droplets to be suspended in a fatty medium, increasing spreadablity. Other molecules, including some phytochemicals, share this dual nature, allowing liaisons between water and fat, and adding these molecules to fats can allow them to harbour water. This changes their properties, for example decreasing their energy value per gram and altering their consistency. However, for a role in fat digestion, cholesterol is not powerful enough to emulsify the fats and so the bile acids come into their own.

When we eat a fatty meal, the mixture that enters our intestines is analysed and bile acids are released according to the fat content to allow the water-based digestive juices to mix with the food and do their job. Once solubilised, and often snipped from the glycerol linker, these hydrophobic molecules are collected by proteins or protein assemblies especially designed to transport them in the blood and prevent the spontaneous clumping of the fats. But even these adaptations are not sufficient to allow the very long fatty acids to be digested by humans. These molecules pass through the gut, becoming fermented en route by the friendly bacteria that live there.

5.3 BODY FAT

At a cellular level, fats form membranes that encase each cell, defining the boundaries of life. The barrier separates the cell and its surroundings, and because the membranes touch water, simple fats will not do. They would be unstable in the watery environment of the body, having the inclination to clump and form droplets rather than form the sheets necessary to encapsulate the cell. To equip lipids for this function, the head is capped with a water-friendly phosphate group to form intuitively named phospholipids. Modified lipids are termed "complex"; simple lipids contain only carbon, hydrogen and a dash of oxygen, but complex lipids are given unique and interesting properties by the addition of other types of atoms including phosphorous, sulfur and nitrogen. The hydrophobic parts shy away from water, while the hydrophilic cap juts out into the watery environment. Mix phospholipids with water and they spontaneously arrange in layers with the hydrophobic parts screened from the aqueous liquid by the phosphate heads. Around cells, sheets composed of two layers of phospholipids assemble so that their hydrophobic fatty parts align on the inside and the outside contains shoulder to shoulder phosphate groups. The arrangement can curve around and rejoin with itself to form bubbles, of which the cell is a natural version. However, these membrane-bound bubbles are not just set aside to compartmentalise cells from the rest of the world. Animals and plants also contain similar boundaries inside the cell-forming pockets, or organelles, that are responsible for different functions: the nucleus is a double phospholipid envelope that houses DNA; vesicles are membrane-bound bubbles formed when food is pulled into the cell from the surroundings; and the mitochondria are also enveloped in a double phospholipid layer to enable them to transform food energy into a form that can be more readily used for cellular processes. The compartments differ from one another in nature and function, but in each case their limits are defined by phospholipid membranes. Proteins lodge within these membranes. These proteins act as gates, so only specific molecules can enter the bubble. They also act as molecular switches, transmitting signals from the outside to the inside: when a signaller such as a hormone binds to the protein switch it induces a structural change that is transmitted through the membrane and into the cell to ensure that all cells are informed of the body's current status.

Fats are made in the endoplasmic reticulum compartment of a cell, an organelle formed from a network of tubes segregated from the main area by a phospholipid membrane. There are two types of endoplasmic reticulum, differentiated according to their appearance under an electron microscope; the first of such images were generated in the 1940s. Before this period, cells were known to contain different organelle compartments, responsible for different functions. The largest of these compartments, the nucleus, was the first organelle to be identified because it was distinguishable with early light microscopes. These microscopes were much like those of the modern day, a tube with lenses that bend the light reflected by an object so that the image is magnified and more details are distinguishable. The power of early microscopes was limited by the ability to carve perfect lenses and align them in the tube. By the end of the 17th century technology was sufficient to allow one to visualise the nuclei of some cells, and the Dutch microscopist Antonie van Leeuwenhoek, now considered the father of microbiology, was one of the first people to see these structures. He worked with salmon blood in which the red blood cells, unlike our equivalents, have nuclei, and what he saw was a "lumen" inside the cells. Without the internet to disseminate his findings, the world remained oblivious for decades to come. Half a century on and Robert Brown, a Scotsman, described the same lumen in more detail and gave it its current name, a name derived from the Latin word meaning kernel. In science, naming a new discovery with your own name leads to eternal fame, but Brown did not opt for such a strategy, possibly because he had already assigned his name to another scientific phenomenon: Brownian motion. Brownian motion is the seemingly random, jerky movement of particles visible under the microscope as smaller invisible particles bump into them. Just as the rediscovery of the nucleus precipitated a wave of scientific advances, so did the discovery of Brownian motion; the principle underpins particle theory and the mathematical model used to describe this motion has been applied far and wide. At the time of naming the nucleus, Brown did not suggest a function and he did not think it was a universal organelle; he had perhaps studied human blood, where the red blood cells shed their nucleus so that they can be packed with more oxygen-carrying haemoglobin. Logically, being human, he assumed that the nucleus in salmon blood was the exception rather than the rule, but in fact most cells have a nucleus.

During the latter half of the 19th century more organelles were identified. As examples, the mitochondria, the power-houses of cells, were described by many microscopists, including Swiss-born Albert Kölliker, and the Golgi apparatus, where the cell processes and packages the proteins and lipids it has made, was identified in 1897 by the Italian physician Camillo Golgi. Although it was clear there were other structures in the cell, the resolution of the light microscope did not allow sufficient characterisation and so the endoplasmic reticulum remained as a shadow until further technological developments occurred, most significantly when microscopy went beyond light and turned to electrons to give increased resolution.

After two centuries of development, light microscopy had discovered its limitation: no matter how powerful the lens, objects cannot be resolved if their size is smaller than the wavelength of light. This problem can be illustrated simply by considering light movement as something like the waves on the ocean. To resolve an object – to see it – its size has to be big enough to disturb the wave as it passes by. Small objects floating under water, like starfish and shells, do not disturb the waves because they are simply too small to have an effect. When a whale swims towards the beach, there comes a point when its shear bulk approaches the size of the waves and causes their patterns to change. We know something is submerged, but only when it stands tall of the wave will we be able to see its form. In the same way, the light waves were sufficiently disturbed by the small structures in the cell for microscopists to know something was there even if they could not resolve its form.

Electrons are part of atoms that whizz so fast it is as though they form a cloud around the atom's core. The clouds are not random, but have tracks or orbitals that the electrons follow. In copper and some other metals, electrons can jump between tracks on adjacent atoms to form the current that we call electricity. These same particles can also be released, forming familiar static electricity. Once free, they can be focused into a beam that moves without the aid of a wire. Beams of electrons move like waves, and their wavelength is 100 000 times smaller than that of light; images made by bending electrons are of much higher resolution than those formed when light shines on the same object.

The first scientists to study cells under the electron microscope, Keith Porter, Albert Claude, and Ernest Fullam, stated, "the electron micrographs disclosed details of cell structure not revealed by other methods of examination".[6] The interior of the cell, in fact, was packed with lace-like structures, which they termed "reticulum", from the Latin word for net. They combined this descriptive noun with a second to describe its place of residence inside the main body of the cell, or plasma, and finally they differentiated the different varieties of this structure: some of the edges of the lace were adorned with tiny beads and others were smooth. The beaded variety they termed rough endoplasmic reticulum and the smoother version just that, smooth endoplasmic reticulum. We now know that, on the first level, rough endoplasmic reticulum is where proteins are made and smooth endoplasmic reticulum is the equivalent for lipids.

The smooth endoplasmic reticulum is also involved in the synthesis and metabolism of steroids and the transportation of newly synthesized proteins to other locations inside or outside the cell. Other functions are specific to cell types; for example, in the case of muscle cells, the smooth endoplasmic reticulum stores calcium for use in muscle contraction, and in the liver, this organelle is used to detoxify drugs and poisons. Of course, under normal circumstances the smooth endoplasmic reticulum is good at its job, but one has to wonder, if this organelle is overworked when producing and processing fat for storage, whether it can perform all the other functions efficiently. If the system is flooded with fat, can the muscles' smooth endoplasmic reticulum still regulate contractions, providing strength without cramps? Can liver cells continue to protect our bodies from cancer-causing poisons? Some people stress their fat processing systems by eating themselves obese within a year.

Finding the organelles was a major scientific step forward. Finding out what the organelles did had to wait. Elucidating the process of fat synthesis, or lipogenesis as it is known in scientific circles, began in the late 1930s with the work of Fritz Lipmann. He was born and educated in Germany, studying both medicine and chemistry before crossing the Atlantic to hold posts at Massachusetts General Hospital and later Harvard Medical School in Boston. Lipmann identified the molecule that provides the carbon atoms to build fatty acids. This species is a bit like a molecular delivery truck that helps the enzyme machines do their jobs. Such helpers are termed co-enzymes because rather than being independent they need to cooperate with enzyme-machines to achieve their goal.

Enzyme-machines are built of protein or occasionally RNA molecules. The machines work on the basis of a simple principle: they alter the rate of chemical reactions by acting as catalysts. Many chemical reactions involve the making or breaking of bonds between atoms, which in turn involves exchange of energy. Enzyme catalysts bind, bend or break the molecules that need to be fused, cleaved or modified. In this way, the progress of the chemical reaction is facilitated, but the enzyme itself is not consumed; at the end of a reaction cycle, the enzyme is exactly as it was at the beginning.

In contrast to enzyme catalysts, co-enzymes are not always regenerated during the cycle of a chemical reaction, often becoming modified or even destroyed. After the reaction, the co-enzyme is regenerated or re-synthesised by other enzyme machines, allowing the catalytic cycle to turn again. Therefore the supply of the co-enzyme can be an important determinant of the rate of the reaction.

The co-enzyme involved in fatty acid synthesis is named co-enzyme A. This molecule is the molecular delivery truck Lipmann discovered. The A in the name stands for "activation of acetate". The loaded truck is distinguished from the empty equivalent by the prefix "acetyl". Lipmann went on to discover much of co-enzyme A's interesting structure. It has a part much like the nucleic acid building blocks of DNA, a part derived from vitamin B_5 (pantothenic acid) and a sticky bit containing a sulfur atom that links up to the acetyl group to allow its transportation and delivery in the cell. The acetyl group is a two carbon block and we now know that fatty acids are made by tethering these units together sequentially. However, at the time of his discovery, Lipmann was in search of a different goal.

Lipmann was studying co-enzyme A and its acetylated equivalent in an attempt to elucidate energy metabolism in cells, and it is in this light that he was awarded the 1953 Nobel Prize in Physiology and Medicine, jointly with another expatriate German, Hans Krebs. Lipmann received the Prize "for his discovery of co-enzyme A and its importance for intermediary metabolism", whereas Krebs was awarded the Prize "for his discovery of the citric acid cycle".

The citric acid cycle also goes by the name of the Krebs cycle. It is the common end point in the metabolism of different types of food: after various processing steps, fats, carbohydrates and proteins all feed into this cycle. Krebs sketched a series of chemical reactions through which energy is released from our food, beginning and ending with a molecule called citric acid. The atoms in this six carbon structure are shuffled, cleaved and finally bound to oxygen to allow the chemical energy trapped in food to be transferred to energy storage molecules. The overall chemical process is much the same as burning, as noted by the 16th-century scientist and artist Leonardo da Vinci, who compared the burning of food by the body to the burning of wax in a candle. On a chemical basis, the processes are indeed similar, but there are important differences: in a candle – or any fire for that matter – the energy is released in an uncontrolled and violent manner to produce heat and light, but in the body such heat would be fatal, and so the burning process is tightly regulated. As the energy trapped in the relatively stable carbon–carbon bonds of our food is released, it is directly transferred into a different type of bond that is more readily accessible to the workings of the cell. In this way a potentially destructive process supports life.

Arguably, the most well known type of energy storage that is readily accessible to the cell is in the bonds formed when phosphorous atoms share electrons. Not just any arrangement of phosphorous atoms has the right properties to meet the energy needs of the cell, but when they share electrons with oxygen atoms, forming a phosphate group, and are then tethered into a three-phosphate chain, their properties are ideal. The phosphate groups are attached to a molecule known as adenosine to make energy-laden adenosine triphosphate (ATP).

ATP was discovered in the 1930s by the Glaswegian chemist Alexander Todd. Two decades on he made the stuff, and soon afterwards Lipmann demonstrated the pivotal role of this molecule in energy transfer to sustain life. The significance of Todd's work was prestigiously acknowledged with a knighthood and the Nobel Prize in chemistry "for his work on nucleotides and nucleotide co-enzymes".

ATP is made by modifying one of the nucleotide building blocks used to make DNA. The actual block is called adenosine (A) and the modification is the addition of three (T) energy-laden phosphate groups (P). Following this naming strategy, ADP is the diphosphate derivative. Each phosphate is linked by an energy-laden bond that can be cleaved to power the cell; however, usually the cell cleaves only one bond, releasing the phosphate and ADP to be recycled back into ATP via the citric acid cycle.

Once the ATP has been made in the mitochondrial energy-house, it passes over the membrane via an exchange protein: ATP exits as ADP is drawn back in to be recharged. In this way the relative amounts of the energy currency stay constant on each side of the membrane. However, the recharging process not only requires ADP, equally important is the phosphate that was cleaved to release energy, and the relative concentrations of this ingredient are regulated by several aspects of nutrition, including the amount of phosphate in the diet and the amount of calcium.

ATP is considered as the molecular energy currency of the cell, but there are other nucleotides doing the same job, for example flavine adenine dinucleotide (FAD), which was also first synthesised by Todd, and nicotinamide dinucleotide (NAD), discovered by British biochemists Arthur Harden and William Youndin in 1906. In some steps, the ATP-recharging process is coupled directly to the citric acid cycle; in others this molecule is charged via the second energy storage and transfer molecule, NAD. Like ATP, NAD contains an adenine base with an attached phosphate, but rather than the phosphates being able to form an energy-storing chain, in NAD the phosphate forms a bridge between the adenine block and another block. The second block contains a pentagon-shaped ribose sugar and a nicotinamide unit (which is chemically distinct from nicotine, despite the similar name). The nicotinamide unit has the ability to become charged with energy by the addition of a proton (a hydrogen nucleus). In this way, NAD becomes energy-laden NADH.

In the energy-laden state, NAD acts as a co-enzyme and also an antioxidant. It can also transfer its energy to generate ATP or use this energy directly to power enzyme-catalysed reactions.

Krebs did not include the fine details of the molecular mechanism when he sketched a series of chemical steps by which cells harvest energy from the six-carbon species citric acid followed by its re-synthesis, ending in a final ninth step where acetyl-co-enzyme A tethers its acetyl group onto the end product to regenerate the starting material and create a cycle. The energy-transferring function of the cycle is what interested Krebs and Lipmann, but acetyl-co-enzyme A does more than just keep the fire stoked, it also stockpiles energy.

Acetyl groups can be generated from any food components: fats, proteins or carbohydrates. Food is mashed in the mouth and then chemically digested into smaller and smaller fragments by the enzymes in the saliva. It is then churned in the stomach using acid both to kill bugs and break down food components a little more before the paste is released into the intestines for further break-down and nutrient absorption. In cells, the food components are processed even more. For the energy function, the common fate of carbohydrates, proteins and fats is to become two-carbon acetyl units. In the cells the acetyl units are released to perform many functions. When energy is needed and there is oxygen about, the acetyl groups are fed into the citric acid cycle to produce ATP, and each time an acetyl group is converted to carbon dioxide and water via the citric acid cycle 12 ATPs are produced.

Direct energy release is not the only fate for acetyl units carried by co-enzyme A. These units can be temporarily tethered to proteins to quench their activity and this process helps to regulate the functions of the cell. Through these interactions, the func-tioning of the cell alters according to the available energy – not the energy stored as fat, or the even more readily available glucose, but the amount of acetyl-co-enzyme A.

The co-enzyme can also donate its ward to make fats and carbo-hydrates. If the energy intake is such that all the ATP reserves are full and the citric acid cycle is well stoked, the remaining acetyl fuel is deposited in longer term storage units. Making stocks of car-bohydrates is tricky because the synthesis of these important nutrients in our bodies is clumsy and inefficient. Only a limited amount can be made from non-carbohydrate starting materials. In contrast, fat synthesis is much more flexible. Fats are happy to be synthesised out of any old food that can be used as energy: pro-teins, carbohydrates, or other fats. In fact, fats can be made from any food components that can be used to generate acetyl-co-enzyme A, which is just about any food that contains calories. Yet despite the body's flexible approach to the starting materials that can be channelled into fat synthesis and the fact that there are organelles dedicated to such processes, the body cannot make all the types of fat needed for health. For energy release, any fat will do, but for the insulation and boundary functions the range is limited and for the signalling functions the nature of the fatty acid is critical. For example, the human body needs long chain fats for the synthesis of docosanoid signallers and these fatty acids are also used in making sperm, the brain and the retina. But size is not everything . . . .

5.4 KINKS

While fatty acid chain length is important, the functions of fats are also linked to the shape of the fatty acid chain: whether it is straight, kinked, twisted or looped. These shapes are a consequence of the nature of carbon atoms, specifically their ability to link to each other in a number of ways. The links that carbon atoms form in fats are called covalent bonds. In such a bond, each atom shares electrons with the neighbouring atom. Bring to mind the previous discussion of electrons and how they like to move around the atom following mathematically predictable paths called orbitals (see The Image of Fat: Body Fat). Electrons are socially inclined and prefer to pair rather than live alone. The theoretical carbon atom has six electrons to neutralise the charge of its six protons lodged in the nucleus. The electrons are ordered into orbitals following a model that applies to all atoms. The understanding of atomic structure has advanced since the first model of these particles was proposed by Niels Bohr, a Danish physicist, in 1903. He proposed a simple version where the protons and neutrons were clustered in the centre of the atom and the electrons lay in progressively stacked shells around the outside, similar to planets orbiting a star. He went on justifiably to receive the Nobel Prize in physics for this work, but we now know that these models are an oversimplification of the real situation. This work has since been combined with that of many other great scientists, including Michael Faraday, Max Planck and Albert Einstein, to promote the understanding of electron motion into the realms of quantum mechanics and thereby account for the fact that electrons are so small that, although they are actually particles, they move in waves like light. For the purpose of this discussion, however, Bohr's first model will do. If one could take an isolated atom of carbon that does not carry charge, it would have six electrons to balance its six protons. Two of these electrons always form a pair and are lodged in the inner shell, but the other four will be alone, each located in a non-overlapping sub-shell – an orbital – on the outer layer. Because of the architecture of these shells, despite carbon having an even number of electrons they cannot pair together. So a carbon atom will naturally form links to other atoms, and such is the strength of the desire of electrons to pair that carbon atoms are never found alone. Covalent bonds are formed by sharing lone electrons with other atoms to create a pair. When two carbon atoms form a single bond, an electron from one carbon pairs and shares the orbital of an electron from the second atom, and because carbon has four single electrons it can create four covalent bonds.

To form a nice regular zigzag pattern along the acyl chain of a fatty acid, all the carbons must link together in the same manner using one of their four electrons to link to the carbon on the left, one to link to the carbon on the right and the remaining two electrons to catch hydrogen atoms. Hydrogen only has one electron and so can form only one bond, which stops the chain branching. When each carbon within the chain grabs two hydrogen atoms, the arrangement results in the carbon chain carrying the maximum number of hydrogen atoms possible and we term the acyl tail "saturated" – saturated with hydrogen.

There is another interesting feature of carbon atoms: when two bond together, they do not have to form one link with just one electron, they can share two or even three electrons each. The type of linkage is named according to the number of electron pairs shared: one electron from each atom means it is a single bond; when two electrons from each atom are involved the link formed is termed a double bond; and when the atoms donate three electrons each it is termed triple. In natural fats, triple bonds are rare, partly because they are less stable than the other types of linkage. Each carbon atom in a double bond shares two electrons with another, leaving two electrons free to make other bonds: one to grab on to the next carbon in the chain and one electron to grab a single hydrogen atom. Linking together to form multiple bonds reduces the amount of hydrogen that can be carried by the chain; the potential to add more without breaking the chain is why these fatty acids are termed unsaturated. In order to achieve double bonds in nature, the carbon rotates against that nice zigzag pattern characteristic of these acyl tails, causing a kink. A chain with just one kink is called monounsaturated, but fats can have many kinks and such chains are called polyunsaturated.

In addition to looking different, at the atomic level kinks give the fat different bulk properties, some of which are visible with the naked eye. For example, the temperature at which the pure fatty acids change from liquid oil to solid fat and back again is affected by kinks. In the case of stearic acid, which contains an 18-carbon chain with no kinks, it melts to form a liquid at around 70 °C. On a warm day it needs a lot of mechanical softening before it can be spread on bread. Having a melting temperature well above body temperature, which usually falls between 36 and 38 °C, also means that this fatty acid readily forms immobile fat deposits in the body, which can be either useful energy stores or disease-causing droplets. This property may be why many of the fat stores in both our bodies and those of animals contain a lot of this fatty acid: it stays put at body temperature. Indeed, its name comes from the Greek word for tallow, which was one of the materials from which this fatty acid was first isolated.

Yet take stearic acid and add a double bond, so that the chain kinks in the natural *cis* conformation and the melting temperature plummets to 13 °C; it is an oil at room temperature rather than a spread. And it is also less likely to form those artery-blocking beads.

In the diet, saturated fats, with all the hydrogen they can possibly carry, are linked to heart disease, whereas some unsaturated fats, with kinks, are thought to protect against this malady; hence, they are deemed to be healthier. Polyunsaturated fats, such as those from sunflower or olive oils, have a maximum of six kinks where hydrogen can be added, and are hailed the healthiest fats. The body makes fats saturated; hence, calorie overload is equivalent to eating saturated fat, and the body can modify the zigzags to make them more or less regular in their pattern by introducing or removing carbon double bond kinks. This process is called dehydrogenation when the bond is made and hydrogenation when it is removed. However, the usefulness of fats to the body is not just dependent on how saturated they are, but also on where the kinks lie along the carbon chain.

We humans have the potential to make double bonds in just four places along the acyl chain, using one of a series of enzymes called desaturases. These desaturases pinch a pair of hydrogen atoms from across a carbon–carbon bond, converting a single link to a double bond. Each of the enzymes is specific with respect to the pair of carbons from which it chooses to remove hydrogen, and it identifies the pair by measuring the position from the carboxyl head. The four desaturases in the human body are Δ^4 desaturase, Δ^5 desaturase, Δ^6 desaturase, and Δ^9 desaturase, and they pinch hydrogen atoms from across the bonds between the fourth and fifth, fifth and sixth, sixth and seventh, and ninth and tenth carbons from the head end of the molecule, respectively. The Δ^9 desaturase converts the widely abundant 18-carbon saturated fatty acid stearic acid into oleic acid by introducing a single kink between the ninth and tenth carbons from the carboxylic acid head. Oleic acid is as abundant as its saturated progenitor, and in the TAG form it makes up the bulk of olive oil, pecan oil, peanut oil, chicken fat and lard.

In-house saturated fatty acid synthesis and the subsequent introduction of selected double bonds allows the body to make many of the fatty acids it desires, but not all. The recipes for protein machines that make certain kinks – kinks on which the body depends for life – are not encoded in our genes. The only way to obtain these nutrients is from the diet, and this is why these fatty acids are termed "essential" nutrients. The range of fatty acids that must be supplied in the diet of each animal varies according to its fatty acid-making machinery; for humans there are two.

Confusingly, and perhaps to demonstrate the artificial distance between the two disciplines, nutritionists tend to designate the place of the double bonds from the tail end of the chain, which is the opposite end to where the chemists start to count and the opposite end to that from which the desaturase enzymes measure; either way, the essential fatty acids are the same. For humans, essential fatty acids have double bonds between the third and fourth carbon or the sixth and seventh from the end of the tail, and they are named using the word omega and the number of the first carbon atom in the double bond: omega-3 fatty acids and omega-6 fatty acids are the two groups of essential fatty acids required by the human body. All other fatty acids can be made from other dietary components, including carbohydrates and proteins.

There are three main reasons for deficiencies in the essential fatty acids among the healthy population. The first is fat phobia. Public education campaigns have linked saturated fats to heart disease and the message has hit home so hard that some people embark upon extremely low fat diets, for themselves and for their children.

Too strict a regime can cause deficiencies in the essential fats and fat-soluble nutrients and, ironically, may not protect against heart disease, because in the body any food can be used to make fat and the first types made are saturated. Given that fat-free foods are often unpalatable, commercial appeal is only obtained when fat is replaced by something else: gums and texturisers are one choice, sugars are another. When saturated fatty acids – the so-called bad fats – are replaced with polyunsaturated fatty acids – the good fats – the level of cholesterol being distributed in the body tends to fall, which has been predicted to reduce the risk of chronic heart disease slightly. However, swapping bad fats for carbohydrates does not seem to have much benefit.[7] This lack of effect may be because the sugars flood the system and so are turned into acetyl units and then into fat or cholesterol by the liver – and fats are first made saturated.

It is not only extremely low fat diets that cause deficiencies of the essential fatty acids; low quality diets, rich in fat, can have the same effect. The sources of essential fatty acids, including some types of fish and various seeds, are often a relatively expensive and acquired taste. Without the motivation and finances to broaden the palette, people may eat a large quantity of fatty foods and still suffer from deficiencies of the essential fatty acids.

The final reason for fat deficiencies is a little more convoluted and is a result of the relative amounts of the two essential fatty acids. The omega-3 and omega-6 fatty acid requirement is inter-related: take in a dose of omega-6 and, for good health, it needs to be balanced by a dose of omega-3. These nutrients act as precursors for the synthesis of several signalling molecules, including the prostaglandins, and the effects of the two groups of signallers are antagonistic. Imbalances in the relative amounts of these fatty acids can lead to imbalances in signalling with the same health consequences as deficiencies. In the modern diet, omega-6 fatty acids are often more abundant than omega-3, leading to an imbalance or deficiency.

Therefore, the problem with fat in the diet of an increasing proportion of the world's population is that the chemistry of the fats does not match the chemistry of the body. In general, the diets common among the Pacific Islanders, Caribbeans, North Americans, Europeans and Australians contain too much fat for good health, illustrated not just by the incidence of obesity, but also by that of other diseases linked to excess dietary fat: heart disease and some types of cancer are examples. The situation does not improve for people opting for low-fat diets filled with sugar; the rapid release of glucose fuel from food stokes the citric acid cycle and the remainder of the acetyl-co-enzyme A flood feeds into fat synthesis. Worse still, despite internally bathing the body in grease, many people have deficiencies in the essential fats because the fat they eat is of low quality. Given that fat chemistry is linked to the arrangement of the atoms in the molecule and our bodies cannot adjust these arrangements to make all the iterations they need, deficiencies of the essential fatty acids cannot be compensated for by other fats or other nutrients. The importance of fat is not just in quantity, but in quality too.

5.5 TWISTS

In addition to double bonds causing kinks in the acyl chain, they can cause twists. The two atomic arrangements differ in that kinks cause the fatty acid chain to change directions dramatically, whereas after a twist the chain more or less continues along the same path. When fats are made naturally, they cycle through a double bond phase before the chain is chemically ironed out; occasionally the less diligent synthesis of fats by bacteria can leave a twist or two. The natural process may seem unnecessarily elaborate, but is an unavoidable consequence of the chemistry of fats. Fats are made by sequentially tethering the two-carbon acetyl group to the growing acyl chain and so they always end up with an even number of carbon atoms in the tail. Acetyl-co-enzyme A off-loads its acetyl part onto the tail end of the chain to make it grow, but needs molecular glue in the form of carboxylic acid. This glue is the same as that used to stick fatty acids to the glycerol linker; in fact, the first unit to begin the fatty acid chain keeps its carboxylic acid glue for this purpose. With subsequent units, once attached the glue is removed in a three-step process, passing through a double-bonded intermediate. However, unlike the double bonds added in at a later date, this intermediate has a different conformation and, rather than forming a kink, it forms a twist in the chain. Scientific naming distinguishes kinks and twists with the prefixes *cis* and *trans*, respectively. The intermediate *trans* twists are chemically ironed out by a process known as hydrogenation. In this reaction, one link in the double bond is broken and hydrogen atom caps are used to pair the carbons' electrons that were formerly shared, restoring the zigzag pattern of the saturated fatty acid. Once in its regular form, the chain is ready for another two-carbon unit to be added. Thus, in our cells, fats are made saturated, passing through a *trans* fat phase. After synthesis, the body adds kinks as required, but limited by its range of desaturases.

Why does the body remove twists? Given that kinks seem to reduce the power of fats to cause health problems, surely twists are desirable too? Twists certainly give some favourable properties in the kitchen. Recall that without kinks or twists stearic acid is a solid at room and body temperature (see The Image of Fat: Kinks). Add a kink and the melting temperature plummets such that a cup of this fatty acid would be an oil. The oil is easier to combine with other ingredients, but it is still far from the perfect spread. If that kink is exchanged for a twist, the fatty acid has a melting temperature of 44 °C, making it more spreadable at room temperature. Stearic acid is not the only fatty acid seemingly to benefit from a twist or two: twisty fatty acids bake crunchier biscuits and help extend the shelf-life of mayonnaise, for example.

In contrast to Nature's diligent removal of twisted double bonds, reactions in chemical vats are less specific. The process that adds twists to our food is actually hydrogenation, a process designed to remove kinks.

Hydrogenation is one of the most widely used catalysed reactions in industry, including the petrochemical industry, and it is an essential step in the manufacturer of many medicines. It also has future applications, for example, in converting garden waste to liquid fuels. However, the most health-influencing contact we have with the products of hydrogenation reactions is dietary: hydrogenation saved our favourite foods from near certain extinction when the links between saturated fats and heart disease became widely known. The fat problem for top level cooks is quite simple: for generations we have based our recipes around animal fats – goose fat was traditionally used to make those roast potatoes super crispy, lard to give pastries their unique light and flaky texture, and in cream saturated fats make desserts delightful – but when these fats were deemed unhealthy the public began shying away, even if the taste and texture of their food was compromised. But there is another incentive to switch to plant products, and that is cost. Animal fats are often more expensive than plant based alternatives. Still, plant oils are just that – oils. Their very consistency does not allow them to substitute directly for animal fats in all recipes. What is more, plant oils have different cooking properties from animal fats, including the smoking temperature (how hot the fat can become before flavour is compromised). Olive oil will never give goose-fat quality roast potatoes.

Fortunately for the health conscious cook, chemists came to the rescue by applying soap technology to our food. James Boyce was born in 1868 as the first generation of Boyces to be born on US soil after his parents' immigration, his father from bonnie Scotland and his mother from the Emerald Isle, Ireland. He became one of the chemists most loved by the American housewife, although few ever knew his name. Rather, they knew – and greatly appreciated – the products he developed: detergents. Boyce was involved in formulating the ingredients of washing powder, and in the process studied the hydrogenation reaction. He was a resourceful fellow and began to investigate whether he could reduce the cost of detergents and soaps by using cotton seed oil, a discarded by-product of cotton manufacture. In Boyce's era, cotton production was booming in the USA. The British fabric manufacturing process had shifted from Indian imports to resourcing the stronger raw material from plantations in the USA and the Caribbean, such that by the mid-19th century "King Cotton" had become the backbone of the southern American economy. The plants were valued principally for their fibres; the seeds could be pressed, but the yellow-coloured oil had little value.

Boyce may have imagined how he could make use of this supply of low cost oil in soap production, modernising cleaning, yet he is unlikely to have dreamed of the revolution he was about to initiate in food technology. He found that, by applying the hydrogenation process to plant oils, he could create a material with a consistency closer to the animal fats used in manufacturing soap – and the animal fats used in cooking. However, recognition of this second opportunity took the ingenuity of a nation that thinks about its stomachs – the French.

Through the 19th century, there was something of a revolution going on in France to celebrate a new found wealth and status. The Eiffel Tower was a demonstration of this turnabout; yet the celebration extended further than the silhouetted Parisian skyline into kitchens throughout the land. But there were problems. Butter was a symbol of luxury and, because the people felt wealthier, they demanded more. The increased demand was met with concern, not because of the now established link between animal fats and heart disease, but because of the cost. And so the search for butter substitutes began, and it was endorsed by no other than Emperor Louis Napoleon III of France. In 1869, he offered a prize to anyone who could make a low-cost butter substitute for the navy and the poor. Scientists set to work and in that same year the French chemist Hippolyte Mège-Mouriès patented his concept of oleo-margarine, later to be abbreviated to margarine. The name was a derivation of a term introduced by one of France's greatest fat chemists of the 19th century, Michel Eugène Chevreul; he was a fat chemist because of his area of interest and not his body form. It was Chevreul who named the fatty acids stearic acid and oleic acid and, in 1813, he coined the term margaric acid, based on the Greek word for pearl *margarís*, to describe what he thought was an undiscovered fatty acid that formed pearly white deposits in fatty material. The label margarine became immortalised, but by 1853 margaric acid was laid to rest when the German structural chemist Wilhelm Heinrich Heintz discovered that the pearly pieces were not a unique substance, but combinations of stearic acid and palmitic acid. The original margarine was also composed of these fatty acids, which could be why it shared the pearly appearance. Mège-Mouriès made his butter substitute by mixing processed beef tallow with skimmed milk to create an emulsion of saturated fatty acid droplets suspended in milky water. Obviously, this product did not beat butter as far as health goes. In fact, it did not beat butter at all. The consistency was butter-like, but the colour was white, and the taste... different. The tallow-milk margarine was simply not buttery enough to gain commercial success.

Paul Sabatier, a French chemist working in Touleuse, on the other hand, was more successful in creating an acceptable margarine. He took Boyce's work beyond the realms of soap and into the up and coming domain of food technology to begin a revolution. Sabatier applied hydrogenation methods to other plant oils to create solid materials not dissimilar to butter, and for this work he shared the 1912 Nobel Prize in Chemistry. A decade on and the Kentucky-born mechanical engineer Murray Raney had just about refined the fundaments of the hydrogenation reaction

for margarine production to what they are today. His hydrogenation catalyst, Raney nickel, found many uses further afield
than fats, but the timing of the launch into margarine manufacture
was perfect – heart disease was on the rise and animal fats were
held responsible. The combined efforts of these and other scientists
allowed modified plant oils to replace butter, and as these products
contained fewer saturated fats and less cholesterol they were
assumed to be better than animal fats, at least as far as heart health
goes. Their commercial success was extraordinary, but not unexpected: not only were plant products deemed healthier, they were
also cheaper, and, equally importantly, hydrogenated plant oils
offered new cooking properties, including more palatable textures
and a longer shelf-life than the natural products. The glory years
went on for decades – until medicine became molecular. When the
molecular markers of heart disease were identified, changes in their
levels in response to diet were studied. Then came the twist:
hydrogenated fats increased the levels of markers linked to heart
disease. The chemical modification of plant oils in the margarine
manufacturing process introduced new properties that were health-
damaging and these new properties were due to the twists in the
fatty acids – the *trans* fats.

At a macro level, mimicking butter using the hydrogenation
process has a simple and logical foundation: one takes cheaper
plant oils that are laden with kinks and irons out some out by
hydrogenation to create a half-way product with chemically
straighter chains, but not so straight that they are the saturated
variety linked to heart disease. It seemed a win–win situation: a
lower cost and seemingly healthier product – at least until science
advanced some more. The problems began to surface when the
molecular details of the hydrogenation process were examined.

The chemical hydrogenation process, for example involving
Raney's metal catalyst, automatically goes through a series of steps
to add hydrogen across the double bond in a stepwise mechanism.
It removes the *cis* double bond, but can cause twists to be formed.
First, the catalyst binds across the double bond, breaking one
carbon electron pair share and leaving the other intact. In this
process, the carbon electrons from the broken bond cling on to the
metal catalyst. Then, the catalyst takes molecular hydrogen from
the mixture and splits it into two hydrogen atoms while shaking a
carbon electron free such that the carbon electron pairs with an

electron from one of the hydrogen atoms and the catalyst grabs onto the other. From this intermediate step, the reaction can go either way: either the catalyst can unite the carbon and hydrogen electrons joined to itself and break away intact, leaving a fatty acid with one fewer kink, or the process can go back the way it came and the hydrogen molecule can be reformed, the catalyst reformed and the fatty acid remain with its kink – or with a twist. Whether the double bond is twisted or kinked depends on whether the fatty acid remains rigid when attached to the catalyst or whether it rotates around itself. In words the difference seems small, almost subjective; in chemistry, the difference is huge. Biological systems do not use Raney nickel metal catalysts. Their catalysts are enzyme machines. They still use metals to catalyse the reaction (which is why our list of essential nutrients includes metals), but rather than being chunks, individual atoms are held in a protein architecture that is far beyond that of Raney's catalysts. The architecture of the metal-containing enzymes is such that when they bind fatty acids to hydrogenate (or dehydrogenate) them, the acyl chains are held rigid and unable to rotate. Because of the sophistication of the catalyst, almost all fatty acids in nature are in the kinky *cis* formation. The exceptions to this rule occur transiently during synthesis and when fatty acids are fermented in the gut; occasionally these fatty acids include a *trans* orientation, but it is only occasional. In most cases, living systems actively remove the twists formed. So, when hydrogenation was applied to food technology there were two revolutions: a visible revolution in the marketplace, incorporating these products into many manufactured foods, and a molecular revolution in our bodies as they tried to accommodate these newcomers. Ironically, this latter revolution has possibly had as significant an effect on the nation's health as the saturated fatty acids and cholesterol these products were designed to replace.

5.6 LOOPS AND RINGS

Chevreul contributed to science at many levels. He elucidated the molecular nature of soap, solid animal and vegetable fats, and oils. His work helped develop many areas of manufacturing, including detergents and candle making. But, perhaps surprisingly for a scientist, he also made significant contributions to the fields of art and religion. In the case of art, he revolutionised colour coding by introducing the concept of simultaneous contrast; the phenomenon that explains why colours can look different according to adjacent colours. In religion he set to work to undermine the growing area of spiritualism. However, despite gaining interdisciplinary fame, arguably his most well known discovery is in science: cholesterol.

Fatty acids are long chains of carbon atoms laced with a hydrogen trim, but some fats form more complex structures, including loops and rings. Cholesterol comes into this category. It does not just have one loop but four: three are made of six carbon atoms and the fourth from five. Some of the carbon atoms are shared between loops, such that they form terraced structures. Cholesterol also has a single hydroxyl group linked onto one side of its rings and a hydrocarbon tail on the other, giving it a dual nature: although predominantly hydrophobic, it can interact with water. Our bodies can synthesise cholesterol, starting with much the same materials as in fat synthesis. Acetyl-co-enzyme A donates its acetyl group to a related charged co-enzyme called acetoacetyl-co-enzyme A to form the dauntingly named hydroxy-methylglutaryl-co-enzyme A, a molecular truck carrying a six-carbon load. This intermediate is tethered to some active phosphate groups before the final carbon atom is snipped off to create a double bond between the now final two carbons in the chain. In some cases, the bond is shunted down the molecule, in others it is left where it landed. Both products are needed to make cholesterol: three of these five carbon units are linked together, one still attached to the active phosphate anchor, the others losing it when they link on. After this, all anchors are cut loose as two of these fifteen carbon intermediates link up. The resultant thirty-carbon structure is largely linear with a few branches. So the next step in cholesterol synthesis is to loop the chain around itself. All this effort in synthesis is not just so we can have heart disease. Cholesterol has some important functions in the body including regulating cell membrane fluidity and acting as a precursor for steroid signaller synthesis: cholesterol is a precursor of vitamin D, sex steroids and the bile acids.

Cholesterol's links to disease are longstanding. It was first harvested from gallstones by the Lyon-born doctor François Poulletier de la Salle in 1769. Yet this deadly fat was not named for nearly 50 years. At first it went by the name cholesterine, given on its rediscovery by Chevreul himself. The naming strategy was simple: an amalgamation of the Greek words for bile (*chole*) and solid (*stereos*) – just what you would expect for a solid discovered within the bile held in the gallbladder. It morphed from the original cholesterine to cholesterol when scientists discovered the hydroxyl moiety attached to one of the loops. Hydroxyl groups, including this one, are characteristic of the chemical family alcohols and, according to chemical naming traditions, all alcohols must end with the chemical suffix *-ol*; think of others, such as methanol or ethanol. The alcohol group is critical for cholesterol's function because it is the only part of the molecule that has polarity, and so when cholesterol slides into the fatty membrane that surrounds the cell, the alcohol group touches the watery environment, correctly positioning the molecule to regulate the fluidity of the membranes.

High levels of cholesterol in the blood are linked to heart disease. The link comes from the observation that the atheromatous plaques that restrict blood flow to the heart contain specific types of fats, including cholesterol. The plaque itself can be divided by chemical composition into three distinct areas. In the centre there is the atheroma, a lump of soft yellowish material that resembles porridge – at least the sort you would expect to find in Greece as its name is literally the Greek term for a lump of gruel. With age, the outer side of the lump becomes hardened with calcium deposits and the artery is less able to do its job, leading to disease. The starting point for this build-up has now been linked to the formation of cholesterol crystals in the artery walls. In addition to circulatory disease, these crystals have also been named and shamed in the development of other degenerative diseases, including Alzheimer's disease. The hardness of these crystals rips the surrounding soft tissues, causing the damage and inflammation that promotes degeneration.

How the cholesterol crystals begin is not clear. Crystals of many substances can be formed in the laboratory. The process forms the foundation of at least two techniques in chemistry. First, through crystals we can obtain a molecule in a very pure form and, second, beaming X-rays through crystals gives a pattern of spots that can be used to determine the atomic arrangement of the molecules that make up the crystals. The second process is dependent on the first. Structural determination is only possible because the atoms or molecules in crystals are arranged in a three-dimensional, orderly, repeating pattern and to obtain such order, the material must be pure. Impurities, essentially other atoms or molecules, disrupt the lattice and eventually the crystal will crumble.

Crystals grow all around us. The most familiar are probably those of water that we call snow or ice, but some rocks are also crystals and so are diamonds. Diamonds are crystals of carbon atoms: eight atoms bond together to form a cube and these cubes pack together to form dense lattices that give the crystals their strength. Cholesterol crystals are diamond-like in many ways. At the macro level, these crystals are clear and beautiful, and at a molecular level, eight molecules form the basic repeating unit of the crystal, but rather than a diamond-like cube, the arrangement is orthorhombic.[8] There are other important differences too: cholesterol crystals are much more fragile than diamonds and their formation is somewhat easier; they seem to precipitate readily, at least in our bodies, whereas growing sizable diamonds still defeats humanity.

The cholesterol crystals are thought to accumulate for a simple reason: too much material gives this molecule too many opportunities to hang out with its kinsfolk and cause trouble. The hydrophobic nature of fats, both the fatty acids and the other hydrophobic materials, means that they cannot be transported around the body without an escort. Alone they would shy away from water-based blood and adhere to any hydrophobic patch they find, for example on an enzyme catalyst or the artery walls, in both cases impairing correct functioning. The body pre-empts the destructive impact of fat on health and packs the blood with fat-sequestering and transport devices, including a protein called albumin, which mops up these and other nutrients that are in excess. Significant amounts of blood fat can be buffered by albumin binding, but when the digestive flood gates are opened after a meal, this protein alone cannot regulate blood fat within the healthy level. Therefore the body employs a special task force called the lipoproteins. There are five types and all but one, the chylomicrons, are named according to the relative amount of proteins they contain. The chylomicrons are the escorts that carry TAGs from the gut to the liver for processing, to the muscles to be burnt as fuel and to the adipose tissue for storage. These escort systems, directly formed from the fat in food, are not the only fat delivery bubbles circulating in the body: the very low density lipoproteins (VLDL) carry home-made fat and cholesterol from the liver to other parts of the body. These bubbles have the lowest protein content of the lipoprotein team. Intermediate density lipoproteins (IDL) are just that – intermediates formed as the VLDL shed their fat load. The two lipoproteins involved in cholesterol transport are low density lipoproteins (LDL), carrying cholesterol from the liver to cells of the body, and the high density lipoprotein (HDL), bringing it back again. The relative amounts of these lipoproteins determine the amount of disease-causing cholesterol in the system and, because of the intertwined nature of fat chemistry, the relative amount of HDL and LDL depends on the flow of fats around the body.

Dietary lipids have been promoted as the source of heart disease for almost a century. These nutrients are absorbed via the gut and include TAGs, phospholipids and cholesterol. Chylomicron assemblies ship these fats around the body from their point of entry. As they circulate they interact with HDL which modifies the function of the chylomicron so that fat metabolism can begin and the fats can be dropped off where needed. Eventually, the chylomicrons become so small that they can be engulfed by the cell, in a process known as endocytosis. The remnants are broken down to release glycerol and fatty acids into the cell, which can be used directly or stored for later. The dependence of chylomicron function on interactions with HDL means that the healthiest fat metabolism is obtained when the amounts of dietary fat are balanced with the amounts synthesised in the liver: the amounts of circulating chylomicron assemblies are balanced by the amount of HDL returning to the liver after delivering their fat load.

HDL starts out as VLDL. These latter assemblies are made up of TAGs and cholesterol synthesised in the liver and are surrounded by protein to aid their transport to other parts of the body. As they circulate, they interact with HDL in the same way as chylomicrons do, loading up with particular proteins that help activate the body to metabolise fat. As fats are delivered the particles shrink to form IDL before returning to the liver to be bled of traces of glycerol and fatty acids, leaving behind LDL charged with relatively high levels of cholesterol. These bubbles are finally engulfed by the liver cells for their protein components to be recycled. This latter point is important: because of the activating interaction of chylomicrons and VLDL with HDL there needs to be a balanced flow of fat, including cholesterol, around the body – balanced with respect to the amount of fat and whether it is the dietary type or the type that is freshly made by our own livers. The relative amounts of these carriers determine health: if more fat is being shipped out from the gut than from the liver, fats are less effectively processed. One may envisage that this inefficiency may help fat stores shrink, but in fact it causes the levels of blood fat to rise, promoting the possibility of cholesterol crystallisation and degenerative disease.

The links between dietary fat and degenerative disease are well publicised. Much less attention is given to the body's own synthesis of fat, which is principally of the saturated type along with cholesterol. This synthesis is an important part of our body's functioning such that even if we do not eat cholesterol-containing foods, the body can make all that it needs and more. The amount of synthesis is strongly dependent on the food we eat. For example, sugars are more readily converted to fats than complex carbohydrates because their rapid digestion floods the body with acetyl-co-enzyme A. Therefore reducing dietary fat to improve health can be counteracted if sugars are used to replace the properties of fats in the kitchen. Various familiar hormones, including insulin, regulate fat synthesis. Insulin activates a protein called lipoprotein lipase (LPL). This protein is an enzyme that chops up TAGs in the chylomicrons and lipoprotein assemblies circulating in the blood. On the insulin signal, LPL in fat cells begins to collect resources from the blood. In contrast, the same signal decreases the function of this protein in muscle tissues, reducing their ability to metabolise fats. Such interrelated regulatory mechanisms have important consequences for health. In a healthy person, a high sugar meal boosts insulin production and, thereby, promotes fat storage rather than fat metabolism, while the calorie excess is converted to saturated fat and cholesterol by the liver. Because of insulin, fat-fear leading to a sugar-loaded diet could aggravate the health problems it was generated to avoid. The significance of the effect depends on the level of health: in some disease states, such as type 2 diabetes, the insulin signal is amplified. The amplification explains the links between insulin resistance and degenerative disease. It is also worth considering whether elevated insulin could also cause body weight gain because of its effects on LPL signalling. These molecular mechanisms suggest that type 2 diabetes is not only caused by unhealthy fat deposits, but it actually causes these fat deposits, creating a cycle of growing girth and impeding weight loss. Given that insulin levels can be raised initially to mask the perturbed glucose tolerance caused by resistance to this signaller, weight gain is one of the earliest symptoms of a degenerative fat chemistry profile, and early detection of glucose intolerance should be a critical consideration for improved healthcare.

How different diets affect cholesterol and, in turn, degenerative disease remains a matter of strong debate. Back at the beginning of the 20th century, animal fats were thought to be the principal problem. Undoubtedly, they provide some cholesterol, but the quantities we eat can be dwarfed by the in-house source. Sheldon Reiser of the United States Department of Agriculture created a stir of fear when he published data suggesting that simple sugars altered the blood lipid profile to promote heart disease.[9,10] His studies showed that rats fed on a super-sweet diet had raised levels of blood fat and LDL cholesterol. It seemed that although some people had embraced nutritional recommendations to reduce meat-fat in an attempt to escape heart disease, their love of soda and ketchup was undoing their good work. Yet despite stimulating enormous media interest, the scientific community remains quiet and sceptical, as usual. Three decades after publication of this work, the waters are beginning to settle and the clearer picture suggests that sugars should not be blamed for cholesterol-linked disease in a healthy person.[11] Reiser's rats were fed such super-sweet diets that the relevance to healthy humans is limited. Yet his studies highlight the complexity of the human body and the fat chemistry within. He demonstrated that what our bodies store is quite different from what we eat. Fat deposits are not just dietary fat, but fat made from other nutrients; cholesterol is not just dietary cholesterol, but also home-made molecules derived from various nutrients, particularly sugar. The amount of home-made fat is dependent on the quality of the diet and also on health: sugar promotes fat deposition via insulin spikes and on a background of insulin resistance the signal is stronger. Taller, Stillman and their successors (and predecessors) who claim that a low-carb diet can help weight loss may be right in some circumstances: a low-sugar diet on a background of pre-diabetes can have some fairly dramatic effects. Banting was an example of the power of dietary manipulation. Cutting back on sugar and adding more vegetables dramatically improved his quality of life. Today there are many people who follow in his footsteps; in some cases, the results of dietary change are as dramatic as medical intervention. It seems that Hippocrates was right when he gave medicinal importance to food.

As our understanding of fat chemistry improves, it seems that dietary saturated fats are not the only cause of heart disease, but the home-made equivalents also play a role, and diet determines the amount of their production. As our understanding of fat chemistry improves, the list of health-damaging charges against *trans* fats grows. It seems that for most of the 20th century the focus on fat was too narrow and foods were designed for ultimate oral pleasure – tasty and appealing – without considering what happens after they are swallowed.

Our understanding of fat chemistry at the beginning of the 20th century was not perfect. After animal fat had been linked to heart disease, fat phobia swept across many countries as, ironically, the incidence of obesity began to rise. There is a link between body weight gain and high-fat diets, but dietary fats are also involved in metabolic protection mechanisms as well as the accumulation of unhealthy fat stores. These protection mechanisms are dependent on the overall diet and also on health, and are critically determined by the plasticity of the fate of acetyl-co-enzyme A. Whether this carbon carrier feeds its fuel into the citric acid cycle to give a virtually instant energy boost or whether it tethers the carbon units together for long-term storage in the form of a fatty acid, or even cholesterol, is determined by fat chemistry; it can also use these units to modify proteins to regulate fat chemistry. On the first level, the fate of acetyl-co-enzyme A is dictated by calories – both those we eat and those we burn. But there is a second level of diet-dependent fat chemistry that goes beyond calories. It is determined, not just by the amount of fat we eat, but also by the quality of fats and the amounts of carbohydrates and proteins, the amounts of minerals and vitamins, micronutrients and even the amount of non-nutrient food components. Chemistry suggests that several changes in the diet over the past few decades are of the type that makes the fat balance promote weight gain.

REFERENCES

1. H. Butler, *Poucher's Perfumes, Cosmetics and Soaps*, Kluwer Academic Publishers, Dordrecht, 10th edn, 2000.
2. M. Best and D. Neuhauser, *Qual. Saf. Health Care*, 2004, **13**, 472–473.
3. S. Vignes and J. Bellanger, *Orphanet J. Rare Dis.*, 2008, **3**, 5–13.

4. C. B. Craig, B. E. Darnell, R. L. Weinsier, M. S. Saag, L. Epps, L. Mullins, W. I. Lapidus, D. M. Ennis, S. S. Akrabawi, P. E. Cornwell and H. E. Sauberlich, *J. Am. Diet. Assoc.*, 1997, **97**, 605–611.

5. H. Takeuchi, S. Sekine, K. Kojima and T. Aoyama, *Asia Pac. J. Clin. Nutr.*, 2008, **17**, 320–323.

6. K. R. Porter, A. Claude and E. F. Fullam, *J. Exp. Med.*, 1945, **81**, 233–246.

7. A. Astrup, J. Dyerberg, P. Elwood, K. Hermansen, F. B. Hu, M. U. Jakobsen, F. J. Kok, R. M. Krauss, J. M. Lecerf, P. LeGrand, P. Nestel, U. Riserus, T. Sanders, A. Sinclair, S. Stender, T. Tholstrup and W. C. Willett, *Am. J. Clin. Nutr.*, 2011, **93**, 684–688.

8. H. S. Shieh, L. G. Hoard and C. E. Nordman, *Nature*, 1977, **267**, 287–289.

9. K. C. Ellwood, M. L. Failla and S. Reiser, *Comp. Biochem. Physiol., A*, 1991, **98**, 323–327.

10. S. Reiser, A. S. Powell, D. J. Scholfield, P. Panda, M. Fields and J. J. Canary, *Am. J. Clin. Nutr.*, 1989, **50**, 1008–1014.

11. S. W. Rizkalla, *Nutr. Met.*, 2010, **7**, 82–99.

CHAPTER 6

Beyond the Helix

Genes were once portrayed as all-powerful, life-determining molecules over which we have no control. With respect to body weight, genetic inheritance – be it in the form of powerful pre-determinants of disease or something like the thrifty genotype – provides a no-blame excuse for being unhealthy; we are what we are and there is little that can be done to alter this fate. However, 21st-century science is knocking genes off their pedestal. We now know that genes define the limits of what we can be, but chemistry determines exactly what we become within these limits. And there is a lot of scope. Chemistry can cause obesity even if the genes say not, and in some cases chemistry can over-ride genetically predestined unhealthy body weight gain. These effects extend from transient adaptation processes that allow survival in a changing environment through to long-lasting chemical memories that are written on genes and can be inherited by the next generation. To understand this science fully, we need to delve into molecular biology.

6.1 MOLECULAR BIOLOGY

Molecular biology is the study of life (biology) down at the smallest units – atoms – and the way they are joined together to form molecules. The term "molecular biology" was first published by William Thomas Astbury in 1951. And the name stuck. After a

Fat Chemistry: The Science behind Obesity
Claire S. Allardyce
© Claire S. Allardyce 2012
Published by the Royal Society of Chemistry, www.rsc.org

decade, even he could not change it; although Astbury proposed the new name of "ultrastructural biology", his peers stuck to the old favourite. And then the field mushroomed into a discipline of its own.

Astbury was actually a physicist infiltrating into biology. Interdisciplinary research is a goldmine for the advancement of science and his contribution was no exception. He applied his knowledge of X-rays to unravel the molecular secrets of proteins and DNA, falling into this field whilst at Leeds University at the time of the height of the wool industry. One of his interests was in examining wool fibres using this technology. From the data he collected he deduced that some of the atoms in the wool proteins (keratin) formed strands that coiled into helical springs.

The first problem with using X-rays to make images of molecules is that, rather than the nice negative images of skeletons we can generate by beaming these rays through our bodies, molecules do not give a photo-quality picture. Actually there is no image at all, just a pretty, spotty pattern called a Patterson function. Scientists have to generate the picture from this pattern using complicated mathematics involving Fourier transforms and phase corrections. Such complex processing was prone to accumulating errors, as in the case of the discovery of thiamine (see Left to our Own Devices: Third Time Lucky). Computers now perform such tasks in a jiffy, but back in Astbury's day the main medium for maths was paper and a pencil. Despite the mathematical challenge, the lure of visualising the molecules of life at an atomic level was exceedingly tantalising and attracted many of the greatest scientific minds of the era.

The second problem with trying to derive a model of a molecule using X-rays is that, even though atoms do diffract X-rays in the same way as much larger objects diffract light, atoms are so small and so diffuse that the intensity of the image is weak. However, if many similar atoms are arranged in a repeating and orderly manner, the diffraction pattern of one combines with that of another until the spots are strong enough to measure accurately. Keratin and DNA naturally form repeating patterns, and so firing X-rays down the length of their fibres allows sufficient amplification to obtain enough data to make a model. But other molecules, for example other proteins and nutrients, are much more asymmetrical in their atomic order and so need to be packed as units into regular arrangements before X-ray diffraction photos can be taken. The ordering usually involves crystallisation, a process that arranges the molecules into beautiful, symmetrical shapes that look much like the cut gemstones arranged in a jeweller's window display; except that these crystals are far too fragile to make ornaments.

In his work, Astbury made major steps forward for science. He predicted the types of regular folding pattern that proteins adopt from the spotty pattern they made. In addition to the coiled springs in keratin, which became known as alpha-helices, in other proteins he found a series of stacked strands, which became known as beta-sheets. However, he did not get as far as making a model of these structural elements; that was the achievement of American-born Linus Pauling who modelled both alpha-helices and beta-sheets[1-3] well before DNA began to reveal its secrets.

As Astbury's work continued, technology for handling the X-rays began to improve to allow progressively clearer spots to be formed, early computers facilitated the maths, and practice and experience proved priceless in preparing the molecules for the experiment. At that time, DNA was known to be made up of sequences of the four nucleotides, but the structure or structures these blocks formed in a molecule was unknown. Astbury made some important advances, and by beaming X-rays along the length of DNA fibres he was able to create spotty patterns that were clear enough to predict that the nucleotide building blocks were "stacked like a pile of plates",[4] but not clear enough to predict the actual structure of the molecule. Astbury was awarded a number of honours for his contribution to science, including a blue plaque on the wall of the house where he lived in Leeds and a research institute bearing his name at the city's University. His inferences formed one of the first steps in the pathway towards the elucidation of the structure of DNA.

It took a woman's touch to obtain X-ray diffraction patterns of DNA of high enough quality to be sure of its structure. The woman was one of the first British molecular biologists, Rosalind Franklin, who, with Raymond Gosling, another member of this distinguished group, provided the critical piece in the DNA structure puzzle: they provided what were described as "amongst the most beautiful X-ray photographs of any substance ever taken".[5] These pictures were not only eye-catching because of their beauty, but also had great significance. By this time the understanding of X-ray patterns had reached such a level that the signs of certain structures could be divined from the spots without need of maths, and Franklin's pictures followed a known pattern. They had no spots near the meridian and this empty space was characteristic of a helix, like that of keratin.

6.2 DNA

The structure of DNA we are most familiar with is a couple of flowing, linear strands wrapped around themselves to form a double helix, but this representation is a gross simplification of the chemical nature of the molecule. The strands represent the molecule's backbone, but rather than simple strings, the real stuff is made up of knobbly chains of alternating links of pentagon-shaped ribose units and smaller, more spherical phosphate units. The backbone does not carry information; rather it positions plate-like beads across the centre of the helix, like a stack of plates, as Astbury predicted. The beads are hooked onto the ribose pentagons. Together, a ribose, a phosphate and a bead are called a nucleotide and these are the basic building blocks of DNA.

DNA is both remarkably refined and extravagant at the same time. An example of the refined nature of DNA is in the code. The order of the four types of nucleotide abbreviated T, A, G, and C, spells out the molecular recipes for life. The recipes are directly transcribed into RNA molecules, some of which function as machines, but the majority are shipped to protein-making stations to be translated into a protein sequence. The translation is based around a triplet code: groups of three nucleotides give instructions for the order of amino acids that appear in the protein chain, and the order is enough to enable chaperone proteins to fold the nascent chain into a complex, yet highly reproducible, three-dimensional molecular structure. The triplet code combined with the four-nucleotide spelling system is refined, representing the minimum code length to provide enough possibilities to define all 22 amino acids used to build proteins, along with triple codes to mark the beginning and the end of the message, without being excessively extravagant. Still, although the triplet code is minimal, once each amino acid has been assigned a code there are some surplus triplet combinations; hence, some amino acids are given more than one code. This multiplicity is a source of variation that does not affect the protein product, but can be traced at a genetic level.

The double helical structure of DNA is elegant, yet becoming extravagant. The Anglo-American team of Francis Crick and James Watson are famed for finally assembling the pieces of the puzzle provided by Astbury, Franklin and others to present a model of the correct structure of this information storage device; a model that, in a grossly simplified form, has become the emblem for many industries and institutions. Part of its popularity may be derived from its fundamental and essential role in life itself, but the human interest in the story of its discovery appeals to the hearts of British people. Although both men went stateside, they made their discovery on British soil and, according to Cambridge folklore, Watson and Crick chose the most English place of thinking to make the announcement of the results of their detective work: their local public house, *The Eagle*. Here, in 1953, they are said to have shared their inferences on the structure of DNA, describing two long helices intertwining. One of the strands is called "sense", because it is indeed the gene, and the other called "antisense", because it is the mould used by the body to make copies of the gene.

DNA strands can act as moulds for genes because of the molecular structure of the plates in the A, T, C and G nucleotides. The C and T plate varieties are smaller than A and G, and a combination of a large and a small plate can exactly straddle the inside of the helix formed by the ribose phosphate backbone. In addition, the A and T plates have two regions of uneven charge distribution – sticky bits – that match to allow two hydrogen bonds (see The Image of Fat) to form between them, and the C and G plates have three. Therefore, the big A of one strand and the small T of the second always pair together, as do the big G and small C. This consistency means that, if you have the mould, you can make the gene by simply matching the pairs. One strand of the double helix can be used to make the other, and this is how DNA is copied when the cell divides. The strands are prised apart and the protein machines match As to Ts and Cs to Gs, and vice versa, to produce two new double helices, each with an original strand and a freshly made complementary equivalent.

The antisense sequence may be helpful, but it is extravagant because some DNA strands stand alone. These single strands include the ends of chromosomes where specialised DNA caps, called telomeres, allow the DNA-replicating machinery to dock. Telomeres consist of several thousand single strand repeats of the simple TTAGGG sequence. Rather than wrapping around a second strand, these regions wrap around themselves to form a stable, stacked arrangement known as a G-quadruplex structure. Other examples of single strands of DNA doing the job of a double strand are found among viruses, for example the smallest virus found in nature, the parvovirus. It gains most fame from causing canine parvovirus and fifth disease. In the former case, the infection triggers vomiting, diarrhoea and immunosuppression that can be fatal in young puppies. The human equivalent is far from fatal; the most noticeable symptom is a fine red rash on the face that looks as though the cheeks have been slapped.

Just as two heads are said to be better than one, it is true that double-stranded DNA genomes do offer advantages over the single-stranded equivalent, including more stability and direct replication of the DNA message, but it does not change the actual information-storage capacity. It is possible to envisage a genome beyond that of a virus that functions adequately using single-stranded DNA molecules and, by default, the double strands are not strictly necessary.

True extravagance in the DNA world seems to come from the amount of non-genetic material that is carried and copied. In addition to those up and running, indispensable genes needed by an organism, vast stretches of what was ambiguously called junk DNA are packed into chromosomes too. Junk DNA gets its name because when it was discovered it seemed to have no purpose. More recent thinking suggests that this junk may be important for regulating gene use, perhaps positioning genes in ways that they are more (or less) accessible to the cell's gene-reading machinery. Junk also reduces the probability that DNA damaging agents, for example environmental chemicals or ultraviolet (UV) light, hit useful genes to cause diseases such as cancer. Given that these roles are far from junk, the name "non-genetic DNA" is now considered more appropriate. The human genome is thought to contain as much as 98% non-genetic DNA, which in some ways helps geneticists, because the search for what makes us what we are needs only to focus on the other 2%. On the other hand, finding this 2% among the so-called junk imposes challenges of its own.

6.3   UNDERPINNING INHERITANCE

Today, we have a mountain of data on human DNA; indeed, we have the entire sequence of the genome. Some parts of this sequence are genes, others seem to regulate genes and the function of other parts is unclear; perhaps it is really junk. In order to communicate findings about different pieces of DNA, scientists have developed an efficient address system based on the location of the gene within the massive twist of the genome. These addresses are known as loci. A locus does not have to be a gene, it can be non-genetic too, but all genes have a locus.

In humans, the genome is divided into 23 chromosomes, little sticks bound together off-centre to give distorted X shapes with two similar short arms and two similar long arms (except in the case of the Y-shaped chromosome in males, where one arm is missing). The chromosomes are numbered according to size, from the largest to the smallest. The gene address is written in a simple code, starting with the number of the chromosome on which the gene resides, then you have to mind your ps and qs, because which letter follows the chromosome number dictates whether the gene locus is on the short or the long arm and, finally, there is a decimal number that indicates how far along that arm the locus lies. Originally, the distance was measured in bands because chromosomes appear to be striped when viewed under the microscope. To find a gene locus, you just counted bands. With the advent of DNA sequencing, more exact addresses can be given by counting the number of nucleotide building blocks along the chromosome. In spite of the added accuracy, many addresses remain in the old fashioned format. Genes are also personalised by their discoverer and given names according to their function. The *ob* gene codes for the appetite regulating protein leptin and the human version is found at 7q31.3 (chromosome 7, long arm, major band 31, sub-band 3).The insulin gene address is 11p15.5 (chromosome 11, short arm, major band 15, sub-band 5).

For gene addresses, the chromosome number is equivalent to the street name and the position on the chromosome is equivalent to the house number. For example, leptin is housed on number seven street. Confusingly for the postman, there are two of nearly every street in the cell: two of each chromosome from 1 to 22, again the exception is chromosome 23 in males where the X is paired with a Y.

The human genome project collected the addresses for all the genes in the human mailing list. This feat was completed in 2003 after 13 years of a challenging, cutting-edge race. With the human genome sequenced, genes can be identified along the length of the DNA helix even if we have no idea what they may do. Given that all genes have the same specific code sequence that is interpreted by DNA reading machinery as a "start" signal – the place where the body needs to begin reading the message the gene encodes – to find genes, all one has to do to find a new gene is scan the sequence for that signal.

The Human Genome Project promised to open a whole world of opportunities to understand ourselves, both now and into the future. It defined the number of genes needed to make a human. Some of these genes affect physical features and others are involved in making life at a molecular level. The project also increased our understanding of genetic variety and the role of some varieties of different genes in disease. In this way, mapping our individual inheritance at a molecular level could give us an insight into what we will become and what diseases will surface. For the future, new technological advances may offer personalised sequencing, but such a goal is currently outside the reach of the average person. When it becomes possible, it will be the first step towards realising Little's dream of personalised medicine at a molecular level; the second step is the arduous task of deciphering exactly what each gene does.

Deciphering gene function starts with sequencing the gene and goes on to translate the information into the molecular instructions. The discovery of the leptin gene allowed its protein product to be made artificially and more of its molecular secrets to be revealed, but its function was already known from the consequences of inheriting non-functional alleles. Working in reverse – starting with a gene and then identifying its function – is more challenging because many genes have overlapping and compensatory functions. The natural process of inheriting non-functional genes can be mimicked in the laboratory and, just as in the case of the ob mouse, these studies can be central to unravelling the function of genes. The artificial equivalent to the natural process is known as knock-out technology. In such studies, a gene's contribution to an individual is abolished – knocked out – and the consequences are recorded. Despite its wide application and significant contribution to understanding gene function, the results of knock-out studies can be misleading, because when one gene is knocked out, others can modulate their function to compensate and the system may adapt to the challenge of being a gene down. Appetite regulation is particularly vulnerable to such compensations. In some cases, but not all, the functions of proteins encoded by different genes overlap so when one ceases to work, another can fill the gap.

Nature's own version of knock-out experiments involves spontaneous errors in matching those A, T, C and G pairs. The copying process is usually highly reproducible and involves proof-reading mechanisms and failsafes that strive to ensure that the genes retain their information. Still, mistakes happen. Carcinogens and mutagens accelerate the rate of these mistakes. And mistakes can completely change the function of the protein product. The spelling mistakes are polymorphisms, whether a single substitution of a nucleotide (point mutation) or the deletion of a whole chunk of DNA. These changes can, in turn, cause a range of symptoms from healthy variety within the population through to unavoidable death. In the case of peas, different spellings of the seed shape gene can cause a round or wrinkled profile and different spellings of the height gene dictate a tall or short stature. Every plant carries seed shape and height genes, but the exact code determines what they will look like. Similarly, every one of us carries a leptin gene, but the exact DNA code of that gene – the allele – influences our body weight. One allele may be super and the other quite ordinary, and when the function is completely abolished disease can result.

Pairing genes offers some protection against disease: if a null gene is inherited, its partner may be able to compensate. Although not an example of disease, consider plant height: if the allele that dictates a tall stature is in fact a growth hormone and the short equivalent is caused by a null gene, one could imagine that when either one or two copies of the functional gene are enough to saturate the signalling pathway, inheriting at least one tall allele will give the characteristic tall height. Only when both gene sites carry a null allele will the plant be short. In this way the tall feature is dominant and short is recessive because of the function of the protein that the gene encodes. One could envisage a similar mechanism in the case of the leptin-null alleles that cause morbid obesity: inheriting a single working copy of this gene is able to compensate for the null allele and so its effect is hidden.

Mendel's pea experiments were at the macro level, studying the whole pea plant for changes in its features. Sweeping generalisations can give a simplistic understanding of a process, but at some point the focus needs to turn to the micro-details. Mendel described genes without understanding molecules; he proposed a theory of inheritance, but his proposition was only verified when it could be mapped at a molecular level. When he described dominant and recessive traits, he did not propose a mechanism for this effect. There are at least two, depending on whether the body uses one or both copies of the gene. One type of dominance is achieved through the effect of the gene product, as in the example of plant height given above, and the other via chemically silencing one member of the gene pair and exclusively resourcing the other. When the gene itself dictates the dominant or recessive nature of inheritance, both genes are used by the body, but one trait masks the other. Another example is at the *ABO* gene site: the *A* and *B* alleles mask the *O* equivalent, but the *O* allele may still be used. In other cases, the body chooses to use one gene and the other is chemically silenced. In peas, the allele for round seed shape is used in preference to the wrinkled variety; one gene is chemically switched off and the other is used exclusively. In this case, the switching off process is determined by the nature of the allele. In other gene pairs, one site is switched off according to which parent donated the gene, and such narrow control can cause disease, such as in the case of Prader–Willi and Angelman's syndromes. Prader–Willi syndrome is linked to a change in the DNA sequence that causes one or more of seven genes on chromosome 15 to be deleted or become defunct, but this genetic change only has an effect on the individual when the altered chromosome is inherited from the paternal line, and when it is, even though working copies of these genes are inherited from Mum, symptoms of disease develop because the body relies on Dad's defunct genes. Mum's contribution is switched off. People with this syndrome are born lethargic and disinterested in the world, including their meals, but manage to overcome these feeding difficulties and usually carry excess body weight in adulthood.

It is not always the paternal genes that are given preference over the maternal equivalents: in Angelman's syndrome the maternally derived genetic material is defunct while the paternal gene is silenced. The result is intellectual and developmental delay, jerky movements and frequent laughter or smiling. In addition to this happy demeanour, patients are prone to piling on the pounds, particularly later in life.

In both Prader–Willi and Angelman's syndromes, the preference for gene silencing is dictated by parental origin and not the function of the allele. These examples of gene silencing operate through an automated mechanism guided by other genes and regions of non-genomic DNA to dictate which gene in each pair is used and which is not. The gene may be marked as it is loaded into the egg or sperm cell, or the markings may be made at conception when the chromosomes can still be linked to their source; either way, once silenced, gene use does not change – not even to alleviate disease symptoms.

There is a dominant form of the peroxisome proliferator-activated receptor gamma (*PPAR-g*) gene that promotes body weight gain. This gene carries the recipe for a protein that forms a fat-sensitive switch. Once activated, it promotes fat storage and glucose metabolism. Despite the problems associated with being too heavy, the dominant allele is used in preference to forms that promote a longer and healthier life because in-built automated programs silence the healthier variety. The actual change in the DNA is small: a C base is replaced by a T – a point mutation.[6] In protein production, the new DNA recipe causes a proline amino acid building block to be replaced with a leucine. This change is significant because proline has a unique property among the amino acids used in protein synthesis in that it causes a kink in the chain. Introducing accidental kinks or ironing out those that are supposed to be there can cause large structural changes that disrupt function. Given that this genetic variant is dominant, just one copy overrides the normal allele, altering brown fat cell morphology so that larger lipid droplets form (see Beyond the Helix: too Much of a Good Thing). These big blobs are less easily metabolised than the more usual equivalents, which impaires the heating process[7] and, thereby, reduces fat burning. Like Mendel's studies of gene pairs in pea plants, the dominant–recessive nature of the *PPAR-g* alleles depends on the sequence of the allele and, therefore, the functionality of the protein produced. However, the effect of inheritance is not due to protein function alone: there is a higher level of complexity that involves chemically modifying the DNA such that the volume control on genes can be altered irrespective of the nature of the allele.

6.4 TO BE OR NOT TO BE

When Mendel studied inheritance he found traits in which one allele was preferred over the other and the effect could not be compensated by other factors. He described only the simplest scenario of recessive–dominant gene pairs and not the complexity of the total jumble of inheritance that makes us all unique. In the case of pea height, there appear to be two possible varieties of the gene: one that makes the plant short and another than makes it grow tall. But outside these famous examples of inheritance, most are more complex. Fortunately, or intuitively, Mendel saw past this confusion and was unperturbed in presenting his theory. He categorised pea height as tall or short. Although there was probably some variation within these two categories, the influence of the short or tall alleles made two distinct groups. Such size distributions are unusual. Species size usually spans a range because the combined contribution of many genes blends together with environmental pressures to determine the trait. Human size is no exception. Combinations of weak and strong gene pairs determine the observed height and we humans have a bell-shaped distribution of height (and weight), with most people having measurements close to the norm, and the further the deviation from the norm the fewer people share the measurement.

The power of genes in determining what we are can be measured and is termed "heritability". In the broad sense, heritability correlates the variation within the phenotype – the traits we see – with the variation within the genotype – genes we inherit; it gauges the contribution of nature versus nurture to what will be. In some cases, measured heritability is close to 100%, such as in Mendel's pea experiments. However, outside the laboratory it is lower; even a pea plant carrying tall genes will be short if nutrients are scarce. In addition to the environment, other genes can skew the heritability value too. When several genes contribute to the same trait, the heritability of any one is altered by the inheritance of others. When one can compensate for the shortfalls of another, heritability can be skewed even further. With such variability, statistically significant data are only achievable with large data sets. Now global communication can be realised with the push of a button, the database has been super-sized. A consortium, aptly named GIANT for Genomewide Investigation of ANthropometric Traits, has been established to probe heritability. GIANT is a data pool from over 200 international institutes that links particular variations within genes to particular traits. Each entry includes the address of the gene and the exact spelling of the allele recipe, coupled to data about that person: height, weight and so on. Using this data pool, researchers can sift through all the numbers and letters to find correlations between alleles and particular features. The benevolent aim of these projects could be to pinpoint markers that determine diseases that lie dormant, such as type 2 diabetes, so that they can be treated before making a permanent mark on health. Yet there is also the possibility of investigating characteristics that make you personally you – in particular weight and height. What GIANT has uncovered is that a giant and extremely complicated web controls inheritance.

The GIANT study suggests that the heritability of height – the amount of variability in height that is caused by genetic factors – is 90%, which means, on average, that the vast majority of the variation in human height is nature; the other one-tenth is determined by nurture. Regulating the heritable component involves at least 180 specific chromosome addresses (loci). This number of loci (either genes, parts of genes or regulatory elements that control genes) is the tip of the iceberg because experts think that it accounts for just 16% of the variation – that is 180 loci for 16% of the difference in height within the population. The loci identified thus far are the strong ones, less able to hide. Researchers are now going after the weaker genetic variants that account for the missing heritability.

It is not only height that has been given the GIANT treatment, but weight too. Using data from a quarter of a million individuals, 14 obesity-related loci have been identified, along with 18 that affect body weight.[8] Most of the loci are linked to the neurological regulation of appetite; others have been identified that may be involved in the regulation of the hypothalamic pathways of energy balance. The combined impact of these genes explains only 1 or 2% of the body weight variance.[9] The discovery that numerous genetic factors are involved in regulating body weight was hardly surprising: weight is critical for survival and yet must be constantly regulated in response to the changing energy demands and fluctuating food supplies. The search for the missing heritability continues.

Yet even without pinpointing candidate genes and other regions of DNA, such as regulatory elements, the heritability of body weight can be estimated based on what we can measure at the whole body level and correlating these data with family trees. As such, body weight heritability is estimated to be between 40 and 70%.[10] Unlike height, there is a heritability range rather than a single value. One possible explanation for the spread is that, whereas height is relatively easy to categorise because its measurement is unambiguous, other traits have a spread distribution as a result of subjective quantification. For example, the heritability of intelligence has been quoted to have about the same range as that of body weight, the exact figure depending on exactly how it is measured. Body weight straddles the categories of easy and difficult to measure. On one hand, the bathroom scales do not lie and give an exact body weight; on the other hand, people tend to be shy of the true reading, introducing error. However, the range may also be due to the different significance of the contributions of environment and genes.

The heritability of body weight can be gauged by studies of identical twin pairs. Even when the individuals are genetically identical, the similarity in the way they gain weight can vary: when the heritability is around 70%, twins, regardless of their environment, gain weight similarly, but when the heritability is around 40%, they don't. One could speculate that those twins who gain weight most similarly have inherited stronger alleles guiding their bodyweight. Indeed, some alleles determine obesity at a level close to totality; for example, the leptin-null gene always predisposes to obesity if inherited in a double dose. However, there is another explanation for these observations that involves environmental influence: the twins that have a similar body weight may do so because of the environment they shared in the womb. There is evidence to suggest that during early developmental stages an individual can be primed to increase the chance of survival in the current environment; first impressions count. The priming involves the same chemical mechanisms that silence recessive genes. Certain genes can be switched off or on according to environmental signals. In this way, priming has effects that are as powerful as inheritance. Examples include preparation for food shortages: if a nutritional shortage is sensed in the womb, the body may be primed for stockpiling in preparation for future famines. The side-effect of such forward planning is the increased chance of unhealthy body weight gain if the famine does not come. Given that measures of heritability do not take priming into account, these annotations can skew the results. Twins who are primed because of their experiences are more likely to have similar body weights than those who did not share such experiences and are living for today. Therefore, heritability includes genes, transient environmental influences and also priming: environmentally induced chemical memories that can last a short time or a lifetime, and in the latter case their contribution to heritability can be easily mistaken for the contribution of genes themselves.

6.5   NATURE VERSUS NURTURE

When James Neel proposed that certain diseases were a con-sequence of natural selection leading to genetic maladaptation, genes were thought to be the ultimate unit of inheritance; the be all and end all. You either inherited a particular allele or you did not. James Watson is amongst several prominent and brilliant scientific minds quoted as suggesting that our fate is largely determined by the DNA code inherited.[11] However, bit by bit a different angle is emerging: heritability is not determined just by our genes, but by how they are used. On one hand, determinants of gene use may be largely written in the genome and follow an automated and immutable pathway. This type of effect certainly seems to be the case for Mendel's dominant–recessive gene pairs: round seeds are dominant over wrinkled regardless of the environment. The wrin-kled seeds are switched off. Priming with the same intractable nature is likely to dictate the altered patterns of gene use that determine the career path of a cell, as well as the more fluid pat-terns that change during growth and development. Even though certain patterns of priming change during growth, the changes and their timing are likely to follow preset and fixed pathways written either in the genes or in non-genetic stretches of DNA. But, on the other hand, there have been several studies, particularly with respect to genes involved in body fat storage and metabolism, that suggest that the priming of certain genes is an adaptive response to the environment, and perhaps once part of the species' survival mechanism; other studies suggest that these responses now cause obesity. The adaptation may still be determined by other genes in the genome, but the impact of the environment can skew measures of heritability. Keith Godfry, the Director of the Centre for Developmental Origins of Health and Disease, University of Southampton, UK, recently brought heritability under the spotlight of 21st-century science with the following statement: "There is considerable evidence that elements of the heritable component of disease susceptibility are transmitted non-genetically, and that environmental influences acting during early development shape disease risk in later life."[12] This means that even if the genetic inheritance does not predestine obesity, environmental triggers in early life can alter gene use to mimic genetically programmed disease.

Godfry was describing an influence on heritability that is not encoded in the DNA sequence as such, but is induced by environmental triggers that alter the way we use the genes we have. Although early examples used to underpin the theory are disputed, new examples are more concrete and a molecular mechanism is beginning to take shape; one that shares some of the molecular details of Mendel's dominant and recessive gene theory and uses the same modifications to switch genes on or off, but not just because of the function of the allele or the sex of the parent from whom it was inherited, but in response to environmental triggers.

It is easy to accept the theory of dominant and recessive genes at a macro level because it fits with what we see in the world around us; at least in some cases. However, at a micro level, credibility is only achievable with a molecular mechanism, and even for the most simple dominant–recessive gene pairs the molecular details are complex. First, the recessive gene must be identified, and then it has to be modified with a degree of reversibility, allowing it to be silenced, yet rekindled under a change of circumstance, for example when the gene is inherited by a new individual or, in some cases, when the environment changes to warrant a different body chemistry profile. The modifications are chemical and can take several forms. Methylation is one example. In this case, a methyl unit (a carbon attached to three hydrogen atoms) is tethered directly onto the DNA's A or C plates, acting as a little hook that holds the two strands more tightly together and, thereby, interferes with the accessibility of the machinery that reads genes. The more methyl hooks on a gene, the less successfully the information is read and the less likely that the gene has an effect. With more and more methyl hooks attached, there comes a point when the DNA reading machine cannot prise the strands apart and the gene is silenced into recessive mode. In this way, the plates not only carry the genetic information, but also a mechanism to control its dissemination.

Scientists can now scan genomes for methyl modifications and this practice has precipitated massive developments in a sub-discipline of genetics called epigenetics. Epigenetics describes all the non-genetic contributions to our inheritance whereby chemical processes adjust the volume of certain genes up or down – even on or off – mimicking the power of genetic inheritance, but it can be reversible. When the volume is turned off, the effect on the individual is the same as inheriting null genes: if the leptin genes are turned off, the individual will become as big as the cousins who inherited a double dose of defunct copies of this gene.

Because only the volume of the genes we inherit is affected, epigenetics is not all-powerful. It is limited by the hand of genes dealt at conception, but it is powerful enough to snuff out the effect of these genes. It cannot add to the genome, but it can take away. It takes away the contribution of genes that could cure disease. There is increasing evidence to suggest that environmentally induced changes to the epigenome take away our innate protection against obesity, and some types of priming can be inherited for several generations. Unsurprisingly, epigenetics is now considered as an important part of the science behind the obesity epidemic.

The description of the environmental impact on epigenetic inheritance has some familiar elements. Jean-Baptiste Lamarck, a French biologist who lived in Lambert's era, proposed a scheme of inheritance where an individual inherited the characteristics its parents used most or acquired during their lifetime. His work was not (and still is not) widely accepted. Natural selection has been given an undisputed molecular basis to drive change in a species, but a similar platform for Lamarckian inheritance has not been discovered; or has it? Epigenetics describes a chemical mechanism by which your life's experiences can be recorded on your genes and inherited by your children. Yet rather than displaying the characteristics their parents used – as Lamarckian inheritance suggests – the experiences have an impact on the child's life in different ways; the parents' actions may prime the body towards particular attributes they never used and a fate they did not share.

Now we know that our fate is not only in our genes but also in the way they are used, which in turn is dependent on our experiences. Genes set the limits – the propensity – of what we can be and our environment and lifestyle pitch the resting place within these limits. In the case of body weight, undoubtedly there is a genetic component: there are a few examples of an all-or-nothing effect, such as the leptin faults that predestine obesity, and some polymorphisms, such as those in the *PPAR-g* or perilipin genes, that make obesity a little bit more or less likely. Undoubtedly there are also experiences that, via epigenetics, make the difference between obese and slim. Imprinting means that any preconceived ideas that heritability is determined solely by the information in the helix need to be shed, and the very real possibility that the apparent heritability of body weight is due to the environmentally induced chemical annotations that alter the way genes are used must be considered. It is also worth considering whether the observations that underpin the thrifty genotype theory are all naturally selected, *i.e.* genetic polymorphisms that give a survival advantage in one environment yet become a disease-causing maladaptation in another, or whether they are due to epigenetically induced changes in the way genes are used that have been triggered in response to much more recent environmental stimuli and, in this case, the obesity epidemic would be preventable.

6.6   THE POWER OF EPIGENETICS

There are other methods of annotating genes in a heritable way apart from DNA methylation. These methods include adding small chemical units to the proteins around which DNA wraps. In a cell, DNA does not stand alone as a linear thread and if it did it would not be stable: its relatively enormous length compared with its width makes it prone to snapping. Therefore the DNA coils around proteins. These proteins were removed when Astbury, Franklin and others collected X-ray diffraction patterns of the molecule, to simplify the maths, but we now know that the helical structure of DNA is just part of a complex coiling and packaging program that both protects the molecule and regulates its use. DNA in its natural, packed form is known as chromatin, and reorganising the packing alters the accessibility of the DNA-reading machinery to genes, giving another mechanism by which epigenetics can regulate gene use.

A third mechanism for altering the influence of gene usage on the phenotype involves RNA molecules. RNA molecules are made by cellular machines using the DNA as a template. Some RNAs, called messenger RNAs, carry recipes for proteins to the protein-making factories in the cell and others ship the amino acid units needed to make the proteins to these same factories. There are also non-coding RNAs, such as microRNAs, that bind to genes or the messenger RNAs to block their use, silencing the contribution of the gene. This type of gene regulation straddles the boundaries between heritable and not: some of the patterns of RNA production are relatively permanent and are inherited by daughter cells, whereas others are much more transient processes that wax and wane with the needs of the cell. Just as for the dominant–recessive gene pairs, the strength of imprinting can be determined by the exact sequence of the allele; some sequences turn the gene off, whilst others just turn the volume down.[13]

Epigenetics is the branch of science dedicated to the study of the combined, chemical, structural and inhibitory silencing mechanisms. These processes are fundamentally important to life as we know it. It is the way that cells remember where they have been and determine what they have become. Nearly all cells in our bodies have exactly the same genes; it is the way they use them that determines whether they are liver cells, brains cells or cancer cells and, therefore, because a liver cell has its job for life, epigenetic markings need to be stable enough to meet this long-term demand. For liver cells to make liver cells, the epigenome needs to be preserved when DNA is replicated as part of the cell division process, when one cell becomes two. Yet there are other instances where changes in the epigenome clearly correspond to nurture: the environment can trigger powerful changes, including changes in the way that we use our genes to cause disease. Take for example Barrett's dysplasia. In this disease, acid reflux into the oesophagus systematically changes the epigenetic markings such that the cells become more like the intestinal cells that are routinely exposed to acid. The protection against cancer is also abolished.[14] The mechanism of altered gene use is likely to occur in a preset and fixed manner, following instructions that are written somewhere in the genome.

There are extreme examples of epigenetic regulation that seem to destroy all understanding of heritability in which the environment – nurture – has a disproportionate effect on traits because of epigenetic imprinting. Such is the case of the wings of the fruitfly *Drosophila melanogaster*. These flies are one of the geneticists' favourite models. There are many strains, including models of human disease and also quirky types that have surprising features. Among this latter group is a strain homozygous for the recessive trait of winglessness, and the heritability is absolute – in some environments. Weather affects the gene use by 100%. Normally, the progeny of the wingless parents follow their genetic destiny and are also wingless. The trait follows Mendelian inheritance and, because winglessness is recessive, there are no hidden alleles that can pop up and give the flies wings; these flies are genetically wingless, through and through. However, epigenetic markings can have a surprising effect: add a little warmth to the nuptials and flies conceived by wingless parents *all* have wings. They inherit alleles that cause winglessness, but heat snuffs out their effect with life-long consequences for anatomy. There are similar examples of temperature-induced changes to the epigenome amongst reptiles. The embryos of tortoises and crocodiles become male or female according to the incubation temperature of the eggs. The temperature triggers imprinting that dictates gender and anatomical changes. Such an influence is powerful and clearly an integral part of species development, driven by genetic programming that responds to an environmental trigger to dictate different patterns of gene use.

Such impressive effects of weather on the epigenome seem to be rare; nevertheless environment can influence body weight via epigenetics. Iceland is not only a hotspot for volcanic activity, it is also a hotspot for human genetic research and obesity is not excluded from such studies. Iceland straddles the tectonic plates of Europe and North America. Its geographical location is closer to Greenland, yet its allegiance lies with Europe. Despite having the ideal location to socialise with either continent, for centuries there has been very little immigration onto the island. Throughout human history, the population has been repeatedly decimated by disease, famine and natural disasters, increasingly streamlining the gene pool. Perhaps to help remember their frequently lost loved ones, the Icelanders have a longstanding passion for genealogy; their extensive and detailed records date back to the first Viking settlers. The results are the closest we are likely to get to

Clarence Little's mouse work in humans: the Icelandic population is one of the most characterised and genetically pure populations of the world.

The epigenetic profile of the Icelanders has intrigued researchers and stirred the world. Among many findings, priming on 13 genes has been statistically correlated with body weight.[15] Several of the genes pinpointed were known to be linked to body weight through previous research. For example, obesity-linked epigenetic markings have been discovered on the *APCDD1* gene, a gene linked by its location to fat deposition in pigs[16] and the percentage of body fat in humans;[17] hence, that the volume control of this gene can be correlated with body weight gain is hardly surprising. The volume control of other genes that encode proteins with functions in feeding regulation and energy balance were also noted to be readjusted among the obese Icelanders.[18] One of these genes, *PRKG1*, encodes the recipe for a protein kinase (PK), specifically the so-called cGMP-dependent protein kinase 1 alpha isozyme. Like all PKs, the *PRKG1* gene product chemically modifies proteins by adding a phosphate chemical group. Protein phosphorylation often works in the same way as DNA methylation: the chemical bumps that are tethered onto these molecules alter their function. The cGMP-dependent PK 1 alpha isozyme adds phosphate bumps onto proteins that regulate foraging behaviour, food acquisition and energy balance, giving a clear mechanism through which changes in the use of this gene could cause changes in body weight.

The profile of gene volume control among the obese Icelanders not only goes some way towards unveiling a mechanism of obesity itself, but also the complications it heralds. The *MMP3* gene, a member of the matrix metalloproteinase (*MMP*) gene family, has been found to be cranked up amongst the obese.[19] The *MMP* genes encode the recipe for matrix metalloproteinase enzymes – molecular scissors. These scissors have zinc atom blades and primarily snip proteins in the liquid that bathes the outside of cells, called the matrix. Collectively, the MMPs can cut all kinds of proteins, but, because the metal blade is lodged inside the sheets and coils of protein structure, each type of scissor specifically snips proteins with a particular shape. The snipping process is a routine part of an ongoing cleaning schedule: continually, proteins are made, used and chopped up again so that the function of the cell adapts to meet its needs and sustain life in a changing environment. Some proteins are degraded quickly and others much more slowly; either way, the degradation prevents the accumulation of outdated signallers and broken or damaged proteins in the cell.

The importance of each MMP can be gauged by the consequences of inheriting a polymorphism that alters its production: using the approach of knock-out studies, natural or otherwise. In the case of MMP3, when the polymorphisms dictate low production, the chance of progressive coronary atherosclerosis and cleft lip/palate is raised,[20] and when MMP3 production is increased, the susceptibility to heart attacks and circulatory problems increases too.[21] Obesity mimics this latter group of polymorphisms by increasing MMP3 levels. This single effect goes some way to explaining the health decline first noted by Hippocrates: obesity alters the volume control on genes to promote heart disease.

Another product of the *MMP* gene family, MMP9, is also overproduced among the obese. This scissor is specially designed to cut the protein collagen. Collagen is the protein-glue that holds the body together. It is made in the usual way from the usual repertoire of 20 amino acids, but after its synthesis two unusual amino acid derivatives are inserted and modified by enzymes that require vitamin C as a cofactor. Because of the ongoing work of MMP9, when collagen synthesis stops, for example because of vitamin C deficiency, body-glue breaks down, leading to scurvy. The reason behind the collagen flux is that only by softening the glue can tissues be remodelled to adapt to the body's needs, and so the MMP9s are important in growth, development, repair and defence.

The process of inflammation is part of the immune response. On a microscopic level, it involves changes in blood flow and the infiltration of the tissue by the defensive white blood cells. On a molecular level, it involves changes in the way genes are used, including increased production of the MMP9 protein. These changes in gene use are a localised, transient and tightly regulated part of routine housekeeping in the body. With more scissors actively snipping the connective collagen glue, it can be softened enough to allow the white blood cells access to the infected site, and remodelling damaged tissue is part of the healing process. Such strategies are also used adaptively for adipose tissue remodelling during weight gain, enabling expansion when fat stores are full.[22]

Among the obese, the *MMP9* gene is used excessively. Usually, there are methyl groups on the gene, but they are shed as body weight rises, leading to the production of more molecular scissors and, in turn, the rate of collagen breakdown is increased, softening tissue structure to allow fat stores to grow and also promoting inflammation. This altered pattern of gene use could generate scurvy symptoms by the opposite mechanism to vitamin C deficiency: rather than blocking collagen synthesis because of a lack of the cofactor, excess MMP9 chops up more collagen than the body is able to replace. Irrespective of the route, the symptoms are the same; vitamin C deficiencies or MMP9 excess can cause inflammation and, eventually, haemorrhage (bleeding), whilst allowing the body to swell.

The epigenetic profile of the obese Icelanders also indicates an altered pattern of *SORCS1* gene imprinting compared with that in the trim. Epigenetic changes in this gene have been linked to the development of diabetes,[23] and genetic polymorphisms that alter the efficiency of this protein have been shown to trigger altered blood glucose levels; the epigenome mimics a genetic disease.[24] Take replication factor C subunit 5 as yet another example of how obesity can cause disease via altered gene use. In response to unhealthy body fat, imprinting of this gene changes to cause the health complications associated with diabetes. Similarly, in the visceral adipose tissue (belly flab), methylation of the *dipeptidyl peptidase-4* (*DPP4*) gene associated with unhealthy weight gain has been linked to the development of insulin resistance and other metabolic disturbances.[25] This is a molecular pathway by which body fat alters gene use to cause the health decline associated with obesity. It is not clear exactly what causes the new epigenome – fat itself or the same trigger that promotes unhealthy body weight, but in the latter case, there is a cautionary tale because the health decline would start before obesity could be diagnosed via the BMI.

Further evidence for the power of epigenetic markings in mimicking genetic disease was unveiled when the epigenomic studies of the heavy Icelanders linked changes in the volume control of the genes involved in the development of syndromes with obesity as a symptom: Prader–Willi, Bardet–Biedl and Wilson–Turner syndromes (the *IL1RAPL2, TTC8* and *DACH2* genes, respectively).[15] In each case, epigenetic markings alter gene use to mimic some of the symptoms of these syndromes. These syndromes are pleiotropic and have many symptoms, and so does obesity. The implications of these findings are wide-reaching: our fate is no longer considered to be written in our genes, but on them as well.

By epigenotyping the Icelanders, researchers have been able to identify changes in the use of genes that could cause obesity, sustain the disease and trigger many of the complications associated with this condition. It is worth reiterating that these epigenetic markings may be a consequence of inheritance. If obesity-priming falls into this category, the epigenetic signature is driven by the inherited DNA sequence in an automated, self-sufficient way, and the difference between the methylation patterns of the obese and the trim would simply be a consequence of different polymorphisms inherited elsewhere in the genome. Our fate would still be written in the genes. The fluidity of the epigenome associated with body fat would follow a set waltz that unfolds during life, with the methyl-footprints marking a pre-determined and perfectly timed choreography along the genome. The steps leading to obesity would be pre-programmed at birth, whether following a direct route, via polymorphisms in the genes linked to appetite regulation and metabolism, or indirect, involving polymorphisms in the sequences that regulate the imprinting of these genes. Genes are back on their pedestal. Yet the obesity epidemic has risen quickly from a static gene pool, suggesting that, rather than nature's fixed response, the epidemic is a result of nurture's call.

Some great scientific minds are beginning to support the idea that the epigenome, at least with respect to body weight, is strongly influenced by nurture and the patterns are freestyle. The epigenetic dance may follow cues written on and in the genes, but it also follows environmental cues and timing. This theory suggests that the environment alone can change the way genes are used to promoted disease. Randy Jirtle of the Department of Radiation Oncology, Duke University Medical Centre, Durham, USA is one of the advocates of this theory. He and his group of researchers have studied the changes in imprinting in response to diet using a particular strain of mice called Agouti. These mice have an unusually active version of the *AgRP* gene: the gene that encodes the agouti protein. Healthy mice produce two pigments for a brown coat colour: dark melanin and yellow phaeomelanin. The brown pigmentation signals are transmitted via the melanocortin receptor 1

(MC1) switch. The agouti protein suppresses pigmentation by binding to and blocking this receptor. Normal amounts of agouti give mice that mousy-brown coat; however, too much agouti protein completely blocks melanin production, allowing the yellow phaeo-melanin pigments to show up. The paler, yellow mouse coat colour is what gives the genetic change its name because it is one of the most easily spotted effects of excessive *AgRP* gene use.

Normally, the agouti protein is only made in the skin, where it acts as a regulator for the synthesis of the melanin pigment. However, when the gene is overactive, colour is not the only phenotype change: production of agouti protein in Agouti mice goes beyond the skin and hair follicles to permeate the entire body. The degree of cross-reactivity among the MC receptors and their various signallers means that over-production of one signaller can cause a range of symptoms by interacting with other receptors. The agouti protein, designed to bind to MC1, can bind to others, including appetite-regulating MC4, preventing correct functioning and causing obesity – Agouti mice are yellow and obese. Specifically, agouti blocks the MC4-receptor in the paraventricular nucleus (PVN), preventing the satiety signalling pathway from being switched on and thereby prolonging the hunger signal so the mice eat themselves obese.

The human equivalent of the *AgRP* gene is named *ASP*.[26] Although the human protein product can block mouse MC1 and MC4 receptors to produce the yellow–obese phenotype, it does not appear to be involved in the regulation of pigmentation in people. It is not even produced in the skin. However, it may be involved in body weight regulation because polymorphisms in the human equivalent of the receptors activated by agouti proteins in mice are linked to body weight in humans. In the human population, variations in the genes that encode the proteins involved in the MC signalling pathway are common, and certain varieties lead to weight gain: 2.5% of the obese population in southern Italy carry characteristic MC4-receptor gene variations that reduce the efficiency of its signalling.[27] In China, 1.5% of the obese population can be accounted for in a similar way, and there are isolated incidences where individual cases of obesity have been attributed to polymorphisms in this receptor.[28] These hardwired pre-determinants of obesity are powerful, and equally powerful effects can be induced by drugs that block the MC4-receptor. These chemicals alter feeding behaviour, knock appetite out of balance with needs and make the body big.[29,30]

Little's first mouse strain – dilute, brown, non-agouti – was free of *AgRP* gene over-expression, giving it both its brown colour and its trim physique. He specifically bred mice to remove the agouti characteristic because this gene is special. Its use is readily influenced by the environment. In fact, *AgRP* gene use is affected by a whole range of environmental factors; it was also the first gene shown to be imprinted in response to nutrition. Jirtle showed that, when nutrient supplements were provided to pregnant Agouti mice, their pups' coats were browner and their weight was lower than when they did not receive the supplements.[31] The nutrients affected the way the mice used their genes, correcting a genetic predisposition towards obesity by epigenetically reducing the use of an overactive gene. The force of a hard-wired disease was overcome by soft wiring. Could this soft wiring be causing the bias in the energy balance among the Maltese, the North Americans and other peoples who are losing the fat war?

The overactive *AgRP* gene is silenced in response to vitamin supplements, and the named vitamins are known to be used by the body in the DNA methylation process. Such supplementation may stimulate methylation in general or it may be that if developing mice do not have enough of these components to switch off all the genes that should be methylated, some may be left on just because of a lack of material. In both cases, the epigenetic priming is responding to an environmental signal and not a genetic one. A genetic response is not unfeasible: the nutritional shortages signal for a different pattern of methylation via the genome as part of an adaptive process to prepare the developing fetus for the world outside. Either way, body weight is influenced by calorie-free nutrition.

The *AgRP* gene in the agouti mice is unusually sensitive to the environment and so when the dietary-induced changes in gene use were first published much of the scientific community remained sceptical. Nevertheless, these findings are evidence to suggest that the diet, or rather the mother's diet, can change the way the body uses its genes and, through this process, change susceptibility to obesity and other diseases.

How important the agouti protein is to the progression of the current obesity epidemic, by blocking the MC4-receptor and quenching its satiety signal, is difficult to gauge. On one hand, mice with an excessively active *AgRP* gene are obese and the human version of this protein can induce the same effect, suggesting that vitamin status could make the difference between fat and slim humans, and the slimming effect of food supplements in these mice suggests that this treatment may help people too. On the other hand, an overactive *AgRP* gene is unusual in both its response to environmental signals and its abundance; after all, how often have you seen a fat, yellow mouse in the wild? So, quite whether this mechanism is of critical importance to human obesity remains to be seen. However, the question over the relevance of the epigenetic regulation of the human equivalent to the agouti protein in explaining the obesity epidemic does not detract from the implications of this finding. Nutrition-induced changes in gene use affect body weight, even if the first gene shown to be controlled in this way is fickle. These findings are too important – too concerning – to ignore.

6.7 SACRIFICES FOR OUR CHILDREN

The environment has a tiered role in determining body weight. The first tier is very obvious, in that the calories in food do contribute to body weight gain. Although we have discussed examples where this contribution seems to be neither direct nor predictable, the first law of thermodynamics assures us that energy cannot be created or destroyed and, therefore, fat – an energy store – cannot be made *de novo*: there must be an input of energy and food is the only route in. And the body cannot shed this extra fuel without transferring the stored energy into another form: movement or heat, as examples. We can be assured that, although the link may not be proportional, there is a link between calories and weight gain. In this first tier, it is worth taking a moment to dwell on the other side of the energy balance because the nature of calorie use determines body weight: diverting food energy to fuel movement or the heating process leaves less to be stored as fat. Balancing what goes in with what goes out has kept generations trim, but the rules no longer seem to apply. Biasing the energy balance is the work of the second tier of the environmental influence on body weight: its effect on fat chemistry. Examples of nutrition, exercise and temperature fall into this tier too, not through the calories they provide or consume, but through their effects on metabolism and appetite regulation that make fat deposition more or less likely. The energy is neither created nor destroyed, but elements of nutrition can dictate the fate of food-energy, and they can also affect appetite. These factors have startling implications: sometimes as powerful as the genetically hardwired causes of obesity; sometimes they even work through genes.

Take, for example, the leptin-producing *ob* gene. Inheriting a double dose of a defective *ob* allele predisposes one to a life of being outsized – grossly outsized. The effect is undeniably impressive. However, the use of leptin genes is modulated by environmental triggers and one of the triggers is your mother's health. The links are convoluted, but sound: Canadian researchers have shown that when Mum has had impaired glucose tolerance during pregnancy, her children use their leptin genes differently.[32] As neonates, they recorded chemical memories of their mum's metabolic problems. These memories promote their own body weight gain later in life.

Impaired glucose tolerance also goes under the alias of pre-diabetes; blood glucose is normal before a meal, but once the digested sugar floods into the body it takes much longer than expected to return to the pre-meal level. Pre-diabetes is a sign that type 2 diabetes is en route. Medical intervention in response to this first signal can delay the onset of the full-blown disease, and one of the medical recommendations to prevent the progression is weight loss. Certainly, shedding pounds in the pre-diabetic state is easier than when the sustained insulin resistance and increased insulin levels associated with type 2 diabetes alter metabolism in favour of weight gain. However, weight loss is also an effective preventative strategy because adiposity alters the way the body uses genes to promote the development of certain diseases, including insulin resistance – the progression of the disease involves an automated positive feedback mechanism.

Insulin resistance can temporarily afflict pregnant women. The changes in body chemistry that occur during pregnancy include the release of signallers that interfere with insulin's message to block the entry of glucose into the mother's cells, thereby forcing it to remain in the bloodstream to feed the growing fetus.[33] The mother's body responds to the lack of glucose in her cells by generating more energy sequestering signals: more insulin. The levels of this hormone are increased in all pregnancies, yet it may be that women with gestational diabetes have reached the limits of their ability to tolerate the rise.

The main hormones thought to be involved in insulin resistance during pregnancy are cortisol and progesterone, but human placental lactogen, prolactin and oestradiol contribute too.[32] Some of these hormones have been implicated in other types of glucose metabolism disorder; some are involved in genetic diseases that lead to obesity. Hence, the reasons for the vulnerability may involve a genetic predisposition. Although such programming is lifelong, cumulative changes may be necessary before there are symptoms, such that the health problems only occur later in life.[34] Certainly, women who experience gestational diabetes are more likely to go on to develop type 2 diabetes, which may only be diagnosed after years of weight gain; the implications of pre-diabetes extend outside the antenatal ward for both mother and child.

If the mother has gestational impaired glucose tolerance, more glucose travels across the placenta to bathe the fetus, increasing growth, but also affecting fat chemistry. These newborns have ongoing high insulin production and susceptibility to low blood glucose levels;[35] they also have altered imprinting that affects the leptin-producing *ob* gene. The fetus was well fed with respect to energy, yet imprinting seems to prime in the direction of stockpiling because the altered leptin production increases the risk of developing obesity and insulin resistance later in life. The inheritance – the genotype – is not changed, but the altered gene use means that the consequence of the inheritance – the phenotype – is not what would be expected. In this way, the changes in gene use that result from nutritional experiences in the womb perturb heritability statistics by mimicking genetic disease, even though the genes say otherwise.

When animals feast on fat their young have a range of symptoms from altered body clocks to altered body weight. Pregnant monkeys fed high-fat diets have young ones born with imprinting that disrupts the use of genes that regulate their body clocks.[36] The same altered gene patterns can be triggered in formerly normal primate youths when they are fed high-fat diets. Fortunately for both groups of monkeys, gene expression reverts to normal with a return to a healthy diet – at least over the timescales studied. Different effects have been studied in rats. When they feast on fat their kittens have altered appetite control and altered food preferences, eating more food than usual and choosing fatty and sugary options.[37] The scenario sounds like the work of galanin, the hunger hormone that stimulates our appetite for fat and sugary foods in response to the same, but the behavioural change occurred over a longer duration than a normal bout of such signalling. Molecular scissors snip circulating galanin so that its concentration falls by one-half in just under four minutes,[38] but the rats' food preferences were examined when they were 76 days old. The long duration of the effect suggests that epigenetics is at play.

Some scientists use the term "malprogramming" to describe cases in which imprinting causes disease. The list of the consequences of malprogramming resulting from a high-fat diet include changes in metabolism, energy balance regulation and the predisposition to mental health-related disorders such as anxiety, depression and attention deficit hyperactivity disorder.[39]

Kevin Grove, of Oregon Health and Science University, Oregon, USA, is one of several researchers who have found scientific evidence to suggest that mental health is intertwined with body weight via diet; something many people knew without data. However, few people would have imagined that the way food changes our mood was more than transient. It seems that these changes may involve both altered sensitivity to hormones and long-lasting changes in gene use. Grove's research interests lie in energy balance and what happens when it goes wrong. His focus is on the regulation of metabolism by the hypothalamus. It seems that the neurocircuitry that is formed *in utero* and shortly after birth has a life-long impact on metabolism and mental health. Grove has a sobering conclusion: "we speculate that future generations will be at increased risk for both metabolic and mental health disorders. Thus, it is critical that future studies identify therapeutic strategies that are effective at preventing maternal [high-fat diet]-induced malprogramming."[39] These researchers speculate that the obesity and anxiety issues are not due to the inheritance of specific genetic polymorphisms that pre-program this state, but from certain genes being activated by the mother's high-fat diet during pregnancy. The sins of the *mothers* shall be visited upon the sons and daughters.

Pharmacological therapies to counterbalance the effects of parental nutrition are decades away, but nutritional manipulation to achieve the same goal is available right now. Such strategies to tackle the obesity epidemic follow the thinking of Peter Kopelman.

He is the Principal of St George's, University of London, and he has dedicated his research career to studying the genetic and endocrinological aspects of obesity. At the start of this millennium, he stated, "rather than aiming for dramatic and expensive interventions that cure obese individuals, inexpensive interventions that have only a small effect on many individuals can nonetheless greatly reduce population-wide financial burden, health risks and impact on quality of life associated with the obesity epidemic."[40] The first step in this programme is to identify the groups of individuals most at risk of unhealthy weight gain, and the children of mothers with a taste for rich foods are one such group. However, the intervention must begin before the child is conceived: young women should be targeted for dietary advice, including information regarding healthy levels of fat in the diet.

Fathers are not innocent of inflicting their children with burdens: the daughters of fat men are more likely to have metabolic disturbances than those with trim fathers, at least amongst rats.[41] The metabolic problems among the daughters of fat dads are similar to those observed to be linked to obesity in the general human population: reduced glucose tolerance, insulin insensitivity and a tendency to stockpile body fat. The girls had this profile before unhealthy body fat had accumulated; they were born this way and the problems worsened with age. The cause was altered gene use. In fact, the use of over 600 genes that are frequently involved in the insulin-producing pancreatic islets was altered, and the use of almost 2500 other genes was also affected. This suggests that the use of one-tenth of our children's genes may be affected by our own body weight. Unhealthy body fat is not a positive influence on the health of the next generation: among the genes affected are those linked to cancer progression and inflammation. Indeed, the gene found to be most up-regulated in response to dad's high-fat diet was *Il13ra2*. Elevated use of this gene is linked to allergies, asthma and allergic lung disease.[42] The increased incidence of these diseases over the past decades is usually attributed to pollution or chemicals in the environment. Now we know that it may be just another complication of fatty food. Clearly, the sins of the fathers are visited on their daughters too.

The consequences of parental diet on their children's health are hard to digest: dietary imbalance or simply a taste for rich foods can have an impact on your child's health, increasing their chance of obesity, diabetes, cancer, inflammatory disease, allergies and mental illness. On the list of sacrifices we are prepared to make for our children's futures, curbing any tendencies to overindulge and controlling our own body weight appears to offer one of the most valuable returns.

6.8   TOO MUCH OF A GOOD THING

Taken together, the work on mice and men explored so far suggests that the effects of parental nutrition on body weight include both nutritional excesses and deficiencies. The excess highlighted at this stage is energy: both fatty foods and fat parents predispose children to a metabolic profile that promotes body weight gain. The deficiencies are those involved in priming: folic acid and vitamin B_{12}. Both vitamins work as co-factors, providing methyl groups for enzymes to tether onto proteins and DNA, but they are not energy providers. These examples are just some of many dietary components that are involved in the second tier of body weight regulation: the tier that does not directly involve calories. Some of the effects are expected, others quite surprising.

Most industrialised countries have a love–hate relationship with caffeine. Let's face it, if we are cola, tea or coffee drinkers, we love the stuff; we lap up reports that suggest coffee drinkers are less likely to develop type 2 diabetes,[43] Parkinson's disease,[44] dementia and Alzheimer's disease,[45] heart rhythm problems, and stroke.[46] Most of these associations are made through surveys, statistically linking this one habit to the incidence of diseases, and because other habits, such as exercise, are not included, many medics take the data with a pinch of salt. Yet there is a chemical foundation to support the health changes linked to caffeine: this chemical does alter the circulation, and most of the diseases it is said to protect against are linked to circulatory malfunctions. Caffeine has other effects on our bodies too; for example, this food component increases the rate of metabolism and fat oxidation.[47] If we are heavy coffee drinkers, we are likely to side with the camp that suggests caffeine can promote weight loss. Yet there is the "hate" side to the relationship with caffeine, and this usually involves people trying to give the stuff up. It is addictive and its consumption is a hard habit to kick. Among those on the caffeine wagon are dieters and pregnant women. Dieters are advised to cut down on caffeine because, despite an increase in the metabolic rate, the overall effect of this food component is thought to be towards body weight gain because it has been shown to impair glucose metabolism,[48] including decreasing insulin sensitivity.[49] With reduced insulin sensitivity comes increased insulin release, which promotes body weight gain by multiple mechanisms. As for pregnant women, this is the group of people that are pressured from every direction when it comes to watching what they eat. My own experience of dietary advice given by the prenatal nurse was along the lines of "now you are pregnant you are free to eat anything, except...". The excluded foods were so encompassing I narrowed the safe foods down to gruel, salt-free of course! Some of the foods that feature on the excluded list are there out of suspicion, but caffeine is named and shamed because of hard evidence: excess caffeine has been linked to causing birth defects, preterm delivery, low birth weight, miscarriage,[50] and heart disease and obesity in the next generation.[51] These latter effects are the results of imprinting a chemical memory on our children's genes: altering their use and thereby altering their fat chemistry and health.

Caffeine's negative role in our development may not come as a surprise; its pick-me-up effect is so powerful that it is familiar and, when an impact on body chemistry can be felt by adults, logical suspicions gather when considering potential harm to a developing fetus. Equally concerning are the impacts of other foods – foods that are not on the excluded list for pregnancy. Many of the entries on this list are unanticipated. Egberto Gaspar de Moura, of the Department of Physiological Sciences, at the State University of Rio de Janeiro, Brazil, demonstrated that thyroid function and body weight of adult animals was affected by maternal nutrition, particularly the amount of protein and energy in the diet.[52] Protein restriction offers protection against unhealthy weight gain by altering thyroid function. The thyroid gland produces the thyroid hormone, which influences the overall rate of metabolism: the thyroid hormone up-regulates metabolism leading to energy wastage, so when production falls or fails, metabolism slows down and energy efficiency increases. An overactive thyroid gland causes patients to be thin and have so much energy they shake, whereas a slow thyroid can lead to weight gain and exhaustion. Moura's protein-restricted rats had altered levels of thyroid hormones: higher levels of triiodothyronine (TI) and lower levels of thyroxin (TY). TY is the precursor of TI; therefore, this altered pattern of thyroid hormones suggests that protein-restriction leads to more processing of TY to the active TI, boosting metabolic rate. The conversion is catalysed by an enzyme known as iodothyronine deiodinase and takes place mostly in the liver. Throughout Moura's study the activity of this liver enzyme was higher in the protein-restricted rat kittens, and this same enzyme was more active in the heat-releasing brown adipose tissue and muscles of their mums during lactation. Thus, protein restriction altered the body chemistry to channel energy into uses other than storage.

At a molecular level, there is a mechanism by which dietary protein can affect thyroid function, and it involves relaying the message of protein status through a series of signallers. The first in the chain is cholecystokinin (CCK). The *chole* part of the name of this signaller is derived from the same origins as the *chole* part of cholesterol: the Greek word for bile. *Cysto* means a sack and *kinin* is derived from the Greek word for movement. Combine all these fragments together and the English translation of CCK is "bile sack mover". And that is what CCK does. It is a peptide signaller responsible for the digestion of fats and proteins that triggers the release of enzymes from the pancreas and bile from the gallbladder into the gut to digest the meal. Despite moving the gallbladder, it has the opposite effect on other areas of the gut, slowing down digestion so that fats can be processed. In addition, CCK suppresses hunger. It is released by the intestines when the chyme that exits the stomach and enters the duodenum is particularly rich in protein or fat. The signal is self-regulatory, triggering the breakdown of the nutrients that caused its release and gradually quenching its own production.

The second signaller in the chain is somatostatin. Like CCK, somatostatin is a peptide hormone made in the gut, but it is also produced in the brain. The two different locations of the hormone give two different functions: regulating the endocrine system and affecting neurotransmission, respectively. In both cases the role of somatostatin is inhibitory. In the endocrine system, it inhibits the release of growth hormone, thyroid stimulating hormone, and gastrointestinal hormones, including those that regulate gut motility and chemical digestion. In fact, somatostatin is a general neuropeptide suppressor. It dulls down the actions of growth hormones, thyroid function and even CCK, the signal that triggered its release.[53] The implications of these wide-reaching actions on our health, and particularly the health of our children, are important: diets rich in protein and fat stimulate the production of CCK, which in turn stimulates the production of the *general* neuropeptide suppressor somatostatin, affecting thyroid function, digestion and pancreatic function. Therefore, protein in our diet is not just a building material for muscles, but a regulatory nutrient that affects the amount of nutrition we absorb from our food and also affects our growth. With respect to these two functions, it works from opposite angles: we feed our children protein to provide enough building blocks for healthy growth, but too much interferes with growth hormone production, and has the opposite effect.

As an explanation for the results of Moura's rat study, the low-protein diet may have reduced the production of CCK, which in turn would have reduced the production of somatostatin. Low somatostatin levels increase neuropeptide signalling, including boosting thyroid and pancreatic functions.[53] Via this molecular mechanism of signalling, it is possible to predict the outcome of the opposite extreme: the rationale predicts that excess protein in the diet could cause changes in metabolism that cause obesity. These pathways suggest that obesity is not only triggered by energy excess, but by excesses of other nutrients too. Today, in many countries around the world, the availability of protein foods is unrestricted and the amount of protein in the diet is, unsurprisingly, high: human beings are genetically programmed to prefer the flavour of protein-rich foods. When we eat them galanin makes us hungry for more. These foods are becoming the most convenient and often the cheapest way of filling our stomachs, appealing to our laziness. Coupled to the psychological element that has us believing that these foods give a health boost irrespective of dietary balance, it is hardly surprising that we tend to overindulge on protein. Given that protein slows down metabolism, this overindulgence could be a possible contributing cause of the obesity epidemic.

Thyroid function is a popular target for anti-obesity drug design, despite the fact that thyroid hormone signalling is not often a cause of obesity. However, via thyroid stimulation, drugs can trigger the channelling of energy in directions other than fat storage. TI is needed by the body to produce uncoupling protein 1

(UP1) – the key protein in heat production in specialised fat cells, called brown adipose cells. Brown adipose cells are indeed coloured brown because they contain so many mitochondria – the energy-producing compartments in cells. These fat cells differ markedly from the white fat cells we love to hate. Indeed, their functions are at the opposite end of the body weight spectrum: white fat cells sequester and store fat in a miserly manner, whereas brown fat cells seize fat solely to fuel their heat-releasing furnaces. Therefore, when fat ends up in white cells, it is there to stay, whereas in brown cells it is actively consumed.

It was the Swiss naturalist Konrad Gessner who first described brown fat, back in 1551.[54] He was born in Zurich, the son of a furrier, which probably gave him ample opportunity to see lard. Later in life, he became an avid documenter and in the course of his mission to catalogue many aspects of the natural world discovered adipose tissue with a difference – rather than lardy white, this fat was brown. He described his findings as "neither fat, nor flesh [*nec pinguitudo, nec caro*] – but something in between". It took another 400 years for the molecular details of fat to be resolved and the surprising accuracy of his description to be proven.

The difference between white and brown fat starts close to their origin, before these fat cells become distinguishable from one another. Fat starts life as dauntingly named mesenchymal stem cells (MSCs). The MSCs have several career choices: they can become bone-building osteoblast cells, cartilage cells or other connective tissue cells, muscle cells or fat cells. Thus, fat and flesh share an origin with bone. What makes fat cells become and stay fat cells is gene use: imprinting determines whether these cells become bone, fat or muscle. Genes marked for use in fat cells include *PPARg*, the molecular switch involved in the regulation of fatty acid storage and glucose metabolism,[55] and use of this gene makes the career choice of cells permanent – it is a sign of commitment to the work of a fat cell, not just during training, but throughout the cell's career.[56] Without PPARg, both brown and white fat cells die. And without fat cells there are metabolic problems because fat has to go somewhere: mice with no PPARg at all are insulin resistant.[56] When there is simply less of this signaller, there is less fat deposition,[57] but not necessarily less fat. The excess causes many health problems; many are similar to those observed in obesity. These findings suggest that the health decline associated with unhealthy body fat stores is not only a function of the stores themselves, but also a consequence of excess fat in general: fat poisoning. Fat cells protect our health by sequestering circulating lipids to allow body chemistry to balance. The amount of PPARg is diet dependent: *trans*-fats reduce PPARg signalling to promote insulin resistance, whereas more natural polyunsaturated fatty acids increase this signalling potential.[58]

Imprinting to direct the fate of fat cells involves many genes: CCAAT-enhancer-binding proteins (C/EBPs) are involved in fat cell maturation, but their effects depend on the colour of the fat,[59] and the confusingly named bone morphogenetic proteins (BMPs) control and support the career choice of both white and brown adipose tissue rather than bone. One, BMP7, singularly promotes the brown career commitment.[60] Brown fat cells use other genes that white fat cells do not, but the key difference between the two fat cell types is the use of *UCP1* genes by brown fat cells.[61] This protein is what allows the brown fat cells to make heat, uncoupling energy storage circuits to keep us warm – and slim. Importantly, if we get cold, really cold, brown fat cells often appear in places where there were formerly only white fat cells,[62] suggesting that the white fat cells could have changed function; their epigenome has altered in response to temperature.

The thyroid function also affects the amount of UP1 in brown fat cells. UP1 is lodged in the membrane surrounding the mito-chondrial compartments and prevents the energy released from food being trapped in the molecular energy currency, ATP. Instead food energy is transformed into heat. UCP1 loosens the tight controls on internal combustion so that metabolism is a little more candle-like than usual. The more heat produced the more energy is used in this process and the less remains for stockpiling. Thus, it follows that encouraging heat production by stimulating the thyroid gland could help weight reduction and control.

Thyroid hormone itself has been trialled for metabolism-boost-ing effects and can induce weight loss. Levothyroxine, a synthetic version of the thyroid hormone that hangs around in the body for much longer than the real thing, is used clinically to address low thyroid function and has the potential for weight management. However, there is a justifiable reluctance to use this therapy because of potential side-effects; thyroid function in the obese is usually normal,[63] suggesting that the production of this hormone is not the key to obesity. Giving the obese more thyroid hormone is effectively mimicking a disease (an overactive thyroid gland) to treat a second disease (obesity) without considering the causes or consequences of either disease.

Whereas drug manipulation of thyroid hormones to control obesity is too risky to implement, dietary manipulation to achieve the same goal is part of our genetically dictated survival mechanism. Rat kittens fed by mothers on a restricted protein diet had a higher level of thyroid function, which means that they produced more thyroid hormones and had a lower body weight than those on energy-restricted diets or those fed conventionally. If this study translates directly to humans, correctly balancing dietary protein may help weight control. In a separate study, Moura showed that after weaning these differences in thyroid hormone levels equilibrate – the detectable chemical memory is erased – but it is possible that during the short period of time when the thyroid levels were altered, further memories were recorded that dictated the future of the body fat.

The role of epigenetics in body weight control shows that, rather than the epidemic of obesity being a result of natural selection and innate genetic inheritance, it can be triggered by environmental changes. These changes can work through genes that have a powerful effect; they can also be annotated on the genes and inherited. And the key environmental triggers are related to nutrition and lifestyle. The importance of these findings suggests that it is time the human race entered the fourth stage of nutritional history, where food has an importance beyond sustenance and the roles of both excesses and deficiencies are considered with equal importance. Food has medicinal status, just as Hippocrates suggested.

REFERENCES

1. L. Pauling and R. B. Corey, *Nature*, 1951, **168**, 550–551.
2. L. Pauling and R. B. Corey, *Proc. Nat. Acad. Sci. USA*, 1951, **37**, 251–256.
3. L. Pauling, R. B. Corey and H. R. Branson, *Proc. Nat. Acad. Sci. USA*, 1951, **37**, 205–211.
4. W. T. Astbury, *Symp. Soc. Exper. Biol.*, 1947, **1**, 66.
5. J. D. Bernal, *Nature*, 1958, **182**, 154.

6. I. Barroso, M. Gurnell, V. E. F. Crowley, M. Agostini, J. W. Schwabe, M. A. Soos, G. L. Maslen, T. D. M. Williams, H. Lewis, A. J. Schafer, V. K. K. Chatterjee and S. O'Rahilly, *Nature*, 1999, **402**, 880–883.
7. S. L. Gray, E. D. Nora, J. Grosse, M. Manieri, T. Stoeger, G. Medina-Gomez, K. Burling, S. Wattler, A. Russ, G. S. H. Yeo, V. K. Chatterjee, S. O'Rahilly, P. J. Voshol, S. Cinti and A. Vidal-Puig, *Diabetes*, 2006, **55**, 2669–2677.
8. E. K. Speliotes, C. J. Willer, S. I. Berndt, K. L. Monda, G. Thorleifsson, A. U. Jackson, H. L. Allen, C. M. Lindgren, J. A. Luan, R. Maegi, J. C. Randall, S. Vedantam, T. W. Winkler, L. Qi, T. Workalemahu, I. M. Heid, V. Steinthorsdottir, H. M. Stringham, M. N. Weedon, E. Wheeler, A. R. Wood, T. Ferreira, R. J. Weyant, A. V. Segre, K. Estrada, L. Liang, J. Nemesh, J.-H. Park, S. Gustafsson, T. O. Kilpelaenen, J. Yang, N. Bouatia-Naji, T. Esko, M. F. Feitosa, Z. Kutalik, M. Mangino, S. Raychaudhuri, A. Scherag, A. V. Smith, R. Welch, J. H. Zhao, K. K. Aben, D. M. Absher, N. Amin, A. L. Dixon, E. Fisher, N. L. Glazer, M. E. Goddard, N. L. Heard-Costa, V. Hoesel, J.-J. Hottenga, A. Johansson, T. Johnson, S. Ketkar, C. Lamina, S. Li, M. F. Moffatt, R. H. Myers, N. Narisu, J. R. B. Perry, M. J. Peters, M. Preuss, S. Ripatti, F. Rivadeneira, C. Sandholt, L. J. Scott, N. J. Timpson, J. P. Tyrer, S. van Wingerden, R. M. Watanabe, C. C. White, F. Wiklund, C. Barlassina, D. I. Chasman, M. N. Cooper, J.-O. Jansson, R. W. Lawrence, N. Pellikka, I. Prokopenko, J. Shi, E. Thiering, H. Alavere, M. T. S. Alibrandi, P. Almgren, A. M. Arnold, T. Aspelund, L. D. Atwood, B. Balkau, A. J. Balmforth, A. J. Bennett, Y. Ben-Shlomo, R. N. Bergman, S. Bergmann, H. Biebermann, A. I. F. Blakemore, T. Boes, L. L. Bonnycastle, S. R. Bornstein, M. J. Brown, T. A. Buchanan, F. Busonero, H. Campbell, F. P. Cappuccio, C. Cavalcanti-Proenca, Y.-D. I. Chen, C.-M. Chen, P. S. Chines, R. Clarke, L. Coin, J. Connell, I. N. M. Day, M. den Heijer, J. Duan, S. Ebrahim, P. Elliott, R. Elosua, G. Eiriksdottir, M. R. Erdos, J. G. Eriksson, M. F. Facheris, S. B. Felix, P. Fischer-Posovszky, A. R. Folsom, N. Friedrich, N. B. Freimer, M. Fu, S. Gaget, P. V. Gejman, E. J. C. Geus, C. Gieger, A. P. Gjesing, A. Goel, P. Goyette, H. Grallert, J. Graessler, D. M. Greenawalt, C. J. Groves, V. Gudnason, C. Guiducci, A.-L.

Hartikainen, N. Hassanali, A. S. Hall, A. S. Havulinna, C. Hayward, A. C. Heath, C. Hengstenberg, A. A. Hicks, A. Hinney, A. Hofman, G. Homuth, J. Hui, W. Igl, C. Iribarren, B. Isomaa, K. B. Jacobs, I. Jarick, E. Jewell, U. John, T. Jorgensen, P. Jousilahti, A. Jula, M. Kaakinen, E. Kajantie, L. M. Kaplan, S. Kathiresan, J. Kettunen, L. Kinnunen, J. W. Knowles, I. Kolcic, I. R. Koenig, S. Koskinen, P. Kovacs, J. Kuusisto, P. Kraft, K. Kvaloy, J. Laitinen, O. Lantieri, C. Lanzani, L. J. Launer, C. Lecoeur, T. Lehtimaeki, G. Lettre, J. Liu, M.-L. Lokki, M. Lorentzon, R. N. Luben, B. Ludwig, P. Manunta, D. Marek, M. Marre, N. G. Martin, W. L. McArdle, A. McCarthy, B. McKnight, T. Meitinger, O. Melander, D. Meyre, K. Midthjell, G. W. Montgomery, M. A. Morken, A. P. Morris, R. Mulic, J. S. Ngwa, M. Nelis, M. J. Neville, D. R. Nyholt, C. J. O'Donnell, S. O'Rahilly, K. K. Ong, B. Oostra, G. Pare, A. N. Parker, M. Perola, I. Pichler, K. H. Pietilaeinen, C. G. P. Platou, O. Polasek, A. Pouta, S. Rafelt, O. Raitakari, N. W. Rayner, M. Ridderstrale, W. Rief, A. Ruokonen, N. R. Robertson, P. Rzehak, V. Salomaa, A. R. Sanders, M. S. Sandhu, S. Sanna, J. Saramies, M. J. Savolainen, S. Scherag, S. Schipf, S. Schreiber, H. Schunkert, K. Silander, J. Sinisalo, D. S. Siscovick, J. H. Smit, N. Soranzo, U. Sovio, J. Stephens, I. Surakka, A. J. Swift, M.-L. Tammesoo, J.-C. Tardif, M. Teder-Laving, T. M. Teslovich, J. R. Thompson, B. Thomson, A. Toenjes, T. Tuomi, J. B. J. van Meurs, G.-J. van Ommen, V. Vatin, J. Viikari, S. Visvikis-Siest, V. Vitart, C. I. G. Vogel, B. F. Voight, L. L. Waite, H. Wallaschofski, G. B. Walters, E. Widen, S. Wiegand, S. H. Wild, G. Willemsen, D. R. Witte, J. C. Witteman, J. Xu, Q. Zhang, L. Zgaga, A. Ziegler, P. Zitting, J. P. Beilby, I. S. Farooqi, J. Hebebrand, H. V. Huikuri, A. L. James, M. Kaehoenen, D. F. Levinson, F. Macciardi, M. S. Nieminen, C. Ohlsson, L. J. Palmer, P. M. Ridker, M. Stumvoll, J. S. Beckmann, H. Boeing, E. Boerwinkle, D. I. Boomsma, M. J. Caulfield, S. J. Chanock, F. S. Collins, L. A. Cupples, G. D. Smith, J. Erdmann, P. Froguel, H. Greonberg, U. Gyllensten, P. Hall, T. Hansen, T. B. Harris, A. T. Hattersley, R. B. Hayes, J. Heinrich, F. B. Hu, K. Hveem, T. Illig, M.-R. Jarvelin, J. Kaprio, F. Karpe, K.-T. Khaw, L. A. Kiemeney, H. Krude, M. Laakso, D. A. Lawlor, A. Metspalu, P. B. Munroe, W. H.

Ouwehand, O. Pedersen, B. W. Penninx, A. Peters, P. P. Pramstaller, T. Quertermous, T. Reinehr, A. Rissanen, I. Rudan, N. J. Samani, P. E. H. Schwarz, A. R. Shuldiner, T. D. Spector, J. Tuomilehto, M. Uda, A. Uitterlinden, T. T. Valle, M. Wabitsch, G. Waeber, N. J. Wareham, H. Watkins, J. F. Wilson, A. F. Wright, M. C. Zillikens, N. Chatterjee, S. A. McCarroll, S. Purcell, E. E. Schadt, P. M. Visscher, T. L. Assimes, I. B. Borecki, P. Deloukas, C. S. Fox, L. C. Groop, T. Haritunians, D. J. Hunter, R. C. Kaplan, K. L. Mohlke, J. R. O'Connell, L. Peltonen, D. Schlessinger, D. P. Strachan, C. M. van Duijn, H. E. Wichmann, T. M. Frayling, U. Thorsteinsdottir, G. R. Abecasis, I. Barroso, M. Boehnke, K. Stefansson, K. E. North, M. I. McCarthy, J. N. Hirschhorn, E. Ingelsson and R. J. F. Loos, *Nat. Gen.*, 2010, **42**, U937–U953.

9. J. Hebebrand, A.-L. Volckmar, N. Knoll and A. Hinney, *Obes. Facts*, 2010, **3**, 294–303.

10. H. H. M. Maes, M. C. Neale and L. J. Eaves, *Behav. Genet.*, 1997, **27**, 600.

11. L. Jaroff, in *Time Magazine*, Time Inc., New York City, 1989, vol. 133.

12. K. M. Godfrey, K. A. Lillycrop, G. C. Burdge, P. D. Gluckman and M. A. Hanson, *Pediatr. Res.*, 2007, **61**, 5R–10R.

13. K. M. Sikora, D. A. Magee, E. W. Berkowicz, D. P. Berry, D. J. Howard, M. P. Mullen, R. D. Evans, D. E. MacHugh and C. Spillane, *BMC Genet.*, 2011, **12**, 4–20.

14. R. C. Fitzgerald, *Gut*, 2009, **58**, 1–2.

15. A. P. Feinberg, R. A. Irizarry, D. Fradin, M. J. Aryee, P. Murakami, T. Aspelund, G. Eiriksdottir, T. B. Harris, L. Launer, V. Gudnason and M. D. Fallin, *Sci. Transl. Med.*, 2010, **2**, 49–67.

16. K. T. Lee, E. W. Park, S. Moon, H. S. Park, H. Y. Kim, G. W. Jang, B. H. Choi, H. Y. Chung, J. W. Lee, I. C. Cheong, S. J. Oh, H. Kim, D. S. Suh and T. H. Kim, *Genomics*, 2006, **87**, 218–224.

17. C. E. Lewis, K. E. North, D. Arnett, I. B. Borecki, H. Coon, R. C. Ellison, S. C. Hunt, A. Oberman, S. S. Rich, M. A. Province and M. B. Miller, *Int. J. Obes.*, 2005, **29**, 639–649.

18. K. R. Kaun and M. B. Sokolowski, *Genome*, 2009, **52**, 1–7.

19. S. Nair, Y. H. Lee, E. Rousseau, M. Cam, P. A. Tataranni, L. J. Baier, C. Bogardus and P. A. Permana, *Diabetologia*, 2005, **48**, 1784–1788.
20. S. Ye, P. Eriksson, A. Hamsten, M. Kurkinen, S. E. Humphries and A. M. Henney, *J. Biol. Chem.*, 1996, **271**, 13055–13060.
21. M. Terashima, H. Akita, K. Kanazawa, N. Inoue, S. Yamada, K. Ito, Y. Matsuda, E. Takai, C. Iwai, H. Kurogane, Y. Yoshida and M. Yokoyama, *Circulation*, 1999, **99**, 2717–2719.
22. A. O'Hara, F.-L. Lim, D. J. Mazzatti and P. Trayhurn, *Pflugers Archiv-European J. Physiol.*, 2009, **458**, 1103–1114.
23. M. O. Goodarzi, D. M. Lehman, K. D. Taylor, X. Guo, J. Cui, M. J. Quinones, S. M. Clee, B. S. Yandell, J. Blangero, W. A. Hsueh, A. D. Attie, M. P. Stern and J. I. Rotter, *Diabetes*, 2007, **56**, 1922–1929.
24. S. M. Clee, B. S. Yandell, K. M. Schueler, M. E. Rabaglia, O. C. Richards, S. M. Raines, E. A. Kabara, D. M. Klass, E. T. K. Mui, D. S. Stapleton, M. P. Gray-Keller, M. B. Young, J. P. Stoehr, H. Lan, I. Boronenkov, P. W. Raess, M. T. Flowers and A. D. Attie, *Nat. Genet.*, 2006, **38**, 688–693.
25. V. Turcot, L. Bouchard, G. Faucher, A. Tchernof, Y. Deshaies, L. Perusse, A. Belisle, S. Marceau, S. Biron, O. Lescelleur, L. Biertho and M.-C. Vohl, *Obesity*, 2011, **19**, 388–395.
26. B. D. Wilson, M. M. Ollmann, L. Kang, M. Stoffel, G. I. Bell and G. S. Barsh, *Hum. Mol. Genet.*, 1995, **4**, 223–230.
27. P. Buono, F. Pasanisi, C. Nardelli, L. Ieno, S. Capone, R. Liguori, C. Finelli, G. Oriani, F. Contaldo and L. Sacchetti, *Clin. Chem.*, 2005, **51**, 1358–1364.
28. C. L. Wang, L. Liang, H. J. Wang, J. F. Fu, J. Hebebrand and A. Hinney, *J. Endocrinol. Invest.*, 2006, **29**, 894–898.
29. W. Fan, B. A. Boston, R. A. Kesterson, V. J. Hurby and R. D. Cone, *Nature*, 1997, **385**, 165–168.
30. D. Huszar, C. A. Lynch, V. Fairchild-Huntress, J. H. Dunmore, Q. Fang, L. R. Berkemeier, W. Gu, R. A. Kesterson, B. A. Boston, R. D. Cone, F. J. Smith, L. A. Campfield, P. Burn and F. Lee, *Cell*, 1997, **88**, 131–141.
31. R. A. Waterland and R. L. Jirtle, *Mol. Cell. Biol.*, 2003, **23**, 5293–5300.

32. L. Bouchard, S. Thibault, S.-P. Guay, M. Santure, A. Monpetit, J. St-Pierre, P. Perron and D. Brisson, *Diabetes Care*, 2010, **33**, 2436–2441.

33. D. B. Carr and S. Gabbe, *Clin. Diabetes*, 1998, **16**, 4–11.

34. T. A. Buchanan and A. H. Xiang, *J. Clin. Invest.*, 2005, **115**, 485–491.

35. L. Kelly, L. Evans and D. Messenger, *Can. Fam. Physician*, 2005, **51**, 688–695.

36. M. Suter, P. Bocock, L. Showalter, M. Hu, C. Shope, R. McKnight, K. Grove, R. Lane and K. Aagaard-Tillery, *FASEB J.*, 2011, **25**, 714–726.

37. T. W. S. Oliveira, C. G. Leandro, T. Deiro, G. D. Perez, D. D. Silva, J. I. Druzian, R. D. Couto and J. M. Barreto-Medeiros, *Lipids*, 2011, **46**, 1071–1074.

38. J. J. Holst, M. Bersani, A. Hvidberg, U. Knigge, E. Christiansen, S. Madsbad, H. Harling and H. Kofod, *Diabetologia*, 1993, **36**, 653–657.

39. E. L. Sullivan, M. S. Smith and K. L. Grove, *Neuroendocrinol.*, 2011, **93**, 1–8.

40. P. G. Kopelman, *Nature*, 2000, **404**, 635–643.

41. S.-F. Ng, R. C. Y. Lin, D. R. Laybutt, R. Barres, J. A. Owens and M. J. Morris, *Nature*, 2010, **467**, U963–U103.

42. M. Wills-Karp, J. Luyimbazi, X. Xu, B. Schofield, T. Y. Neben, C. L. Karp and D. D. Donaldson, *Science*, 1998, **282**, 2258–2261.

43. R. M. van Dam and F. B. Hu, *J. Am. Med. Assoc.*, 2005, **294**, 97–104.

44. A. Ascherio, S. M. M. Zhang, M. A. Hernan, I. Kawachi, G. A. Colditz, F. E. Speizer and W. C. Willett, *Ann. Neurol.*, 2001, **50**, 56–63.

45. M. H. Eskelinen and M. Kivipelto, *J. Alzheimers Dis.*, 2010, **20**, S167–S174.

46. W. Zhang, E. Lopez-Garcia, T. Y. Li, F. B. Hu and R. M. van Dam, *Circulation*, 2009, **119**, E298–E298.

47. R. Hursel, W. Viechtbauer, A. G. Dulloo, A. Tremblay, L. Tappy, W. Rumpler and M. S. Westerterp-Plantenga, *Obes. Rev.*, 2011, **12**, E573–E581.

48. J. D. Lane, C. E. Barkauskas, R. S. Surwit and M. N. Feinglos, *Diabetes Care*, 2004, **27**, 2047–2048.

49. G. B. Keijzers, B. E. De Galan, C. J. Tack and P. Smits, *Diabetes Care*, 2002, **25**, 364–369.
50. N. A. Graham, C. J. Hammond, III and M. S. Gold, *Am. J. Obstet. Gynecol.*, 2008, **199**, 15.
51. C. C. Wendler, M. Busovsky-McNeal, S. Ghatpande, A. Kalinowski, K. S. Russell and S. A. Rivkees, *FASEB J.*, 2009, **23**, 1272–1278.
52. P. C. Lisbo, M. C. F. Passos, S. C. P. Dutra, R. S. Santos, I. T. Bonomo, A. P. Cabanelas, C. C. Pazos-Moura and E. G. Moura, *J. Endocrinol.*, 2003, **177**, 261–267.
53. K. C. K. Lloyd, V. Maxwell, C. N. Chuang, H. C. Wong, A. H. Soll and J. H. Walsh, *Peptides*, 1994, **15**, 223–227.
54. B. Cannon and J. Nedergaard, *Nature*, 2008, **454**, 947–948.
55. S. R. Farmer, *Cell Metab*, 2006, **4**, 263–273.
56. E. D. Rosen and O. A. MacDougald, *Nat. Rev. Mol. Cell Biol.*, 2006, **7**, 885–896.
57. W. M. He, Y. Barak, A. Hevener, P. Olson, D. Liao, J. Le, M. Nelson, E. Ong, J. M. Olefsky and R. M. Evans, *Proc. Nat. Acad. Sci. USA*, 2003, **100**, 15712–15717.
58. K. Reue and J. Phan, *Curr. Opin. Clin. Nutr. Metab. Care*, 2006, **9**, 436–441.
59. H. G. Linhart, K. Ishimura-Oka, F. DeMayo, T. Kibe, D. Repka, B. Poindexter, R. J. Bick and G. J. Darlington, *Proc. Nat. Acad. Sci. USA*, 2001, **98**, 12532–12537.
60. Y.-H. Tseng, E. Kokkotou, T. J. Schulz, T. L. Huang, J. N. Winnay, C. M. Taniguchi, T. T. Tran, R. Suzuki, D. O. Espinoza, Y. Yamamoto, M. J. Ahrens, A. T. Dudley, A. W. Norris, R. N. Kulkarni and C. R. Kahn, *Nature*, 2008, **454**, U1000–U1044.
61. J. Kopecky, M. Baudysova, F. Zanotti, D. Janikova, S. Pavelka and J. Houstek, *J. Biol. Chem.*, 1990, **265**, 22204–22209.
62. S. Cinti, *Prostag Leukotr. EFA*, 2005, **73**, 9–15.
63. A. Strata, G. Ugolotti, C. Contini, G. Magnati, C. Pugnoli, F. Tirelli and U. Zuliani, *Int. J.Obes.*, 1978, **2**, 333–341.

CHAPTER 7

A Question of Dose

One of the problems of many weight control regimes is the focus: calories. They are about 100 years behind scientific advances. The ability of fat chemistry to regulate appetite and metabolism so that body weight is buffered within a healthy range depends on many nutrients. Many nutrients that are important for weight regulation have little or no calorific value.

Nutritional needs are now defined and well publicised. Nearly every food product has some nutritional information printed on the pack. Single numbers are often presented. These are not absolute values, but averages for the adult population. The actual amounts of different nutrients required by the body are individual. Nevertheless, there are some generalisations that can be made, for example some nutrients are needed in greater amounts than others.

7.1 A NUTRITIONAL OVERVIEW

The body needs three categories of food components in the greatest amounts – carbohydrate, fat and protein – and, accordingly, these nutrients are given the prefix macro, derived from the Greek word *macros*, meaning large. In addition, water can be added to this class because, although it is not chemical sustenance as such, large quantities are essential for good health, but too much can cause health problems: people die as a result of drinking too much water, but usually only after water-drinking contests. The body is made

Fat Chemistry: The Science behind Obesity
Claire S. Allardyce
© Claire S. Allardyce 2012
Published by the Royal Society of Chemistry, www.rsc.org

up of between 50 and 70% water, with the difference depending on the amount of fat stored: the more fat the less water and vice versa. The amount of water is in constant flux: it can evaporate, through sweat and on the breath, and is used to flush poisons out in the urine. As these processes are ongoing, lost water needs to be frequently replaced.

Carbohydrate is another macronutrient. It is principally used as an energy source, but also for a variety of functions around the body, including forming identity cards so body parts can be recognised and distinguished. This macronutrient family can be divided into two groups: simple sugars and complex carbohydrates. Although they share the same generic molecular structure, these groups have quite distinct chemical properties in the lab and in the body. In fact, complex carbohydrates are composed of many individual sugar units tethered together in a chain. These larger molecules dissolve less readily in water and so can form structures in living organisms: cellulose – or dietary fibre – is a complex carbohydrate that gives plants their form, and its resistance to human digestion gives our guts form too. In other cases, simple sugars are joined together to form a larger, insoluble molecule that serves as an energy reservoir. For example, muscle cells attach glucose molecules together to form a glycogen reserve to provide the first fuel for movement before fat metabolism begins. Starch is the plant's equivalent to glycogen and provides the healthiest supply of sugar, slowly releasing units to allow the cells to build up reserves of glycogen without being flooded by a sugar-rush. Of course, when the carbohydrates in food are already in the simple sugar form, all the energy arrives at the same time and such flooding channels sugar into fat production. The dose tolerated depends on genes and health: in the case of insulin resistance, the tolerated dose of sugar falls and the recommended daily intake should reflect this change. The quality of carbohydrates ingested affects body weight, with the smaller simple sugars causing hormonal surges that promote weight gain while their complex cousins offer a steady and dependable energy source.

Like carbohydrates, fats are also energy-giving foods, but they have other functions too, for example making skin waterproof and insulating nerves so short circuits do not confuse communication in the brain. Just as with carbohydrates, there is a quality factor in this category of nutrients: some fats are essential; others can be synthesised by the body as required; and yet more – the *trans* fats, for example – could be health-damaging. Different types of fat have different effects on body weight and other aspects of health (see The Image of Fat).

Protein can be used as a source of energy, but only as a last resort because these nutrients are a valuable body building material, and so are preferentially used in maintenance and repair. Protein also acts as a signaller, and the response is determined by both the quantity and quality of this macronutrient. The quantity of protein in the diet determines the release of various regulatory hormones, including somatostatin, the general neurosuppressor (see Beyond the Helix: Too Much of a Good Thing), and the quality of protein – the relative abundance of different amino acid building blocks in the meal – balances the branch of signalling that involves the amino acids glutamate and aspartate (see A question of Dose: Generally Recognised as Safe). In both cases, imbalances have been linked to changes in body weight.

Micronutrients include the essential nutrients the body needs for life, other than carbohydrate, fat, protein and water. The quantity of each micronutrient required for good health is much smaller than that of the four macronutrients, perhaps just a few micrograms (a millionth of a gram) each day or less. Now, do not be misled by the amount because some of these nutrients are powerful: powerful enough to prevent scurvy, pellagra and beriberi amongst other diseases.

Minerals are the most robust micronutrients and include metals, usually in the form of their salts, extracted from rocks and present in the water drunk by both ourselves and our food. These metals have many functions in the body, forming part of some enzyme machines and also in signalling. Non-metal minerals – phosphorus, iodine and sulfur, for example – have similar multi-tiered roles in health. The levels of consumption of several minerals have been linked to obesity. Take, for example, vanadium. When it is found polluting the environment it is considered as a poison, but it is also an important micronutrient needed by the body for good health. Some complexes of this metal increase the amount of cancer-protection molecules in the body,[1] some inhibit tumour cell growth and others help lower blood sugar. The forms of vanadium called vanadate and vanadyl ions are often referred to as "insulin mimics". In fact, although these derivatives lower blood sugar levels, they are not a substitute for the hormone; instead, they make the body more sensitive to its signal. In a sense, vanadium complexes reverse insulin resistance, such that drugs to treat type 2 diabetes are often based on this mineral. It follows that diet could help reduce the symptoms of the disease; or exacerbate the problem. Certainly, nations that feast on foods rich in vanadium – salads, radishes, parsley, lettuce and seafood – tend to have a low incidence of insulin resistance.[2]

Vanadium was first discovered by the Spanish–Mexican scientist Andrés Manuel del Río in 1801. He described a new element that he called erythronium, derived from the Greek word for red because of its colour. Yet his peers were not convinced that this material warranted element status, suggesting it was merely a form of the already known element chromium. Finally, it received its due recognition when it was rediscovered in 1831 by the Swedish chemist Nils Gabriel Sefström, who opted for a more sophisticated naming strategy using the name Vanadís – a goddess of beauty and fertility – for inspiration.

It is interesting that vanadium was first confused with chromium in the laboratory, because there is a parallel in the body: although vanadium works against chromium in some functions, both minerals seem to be involved in regulating blood glucose, and just as vanadium boosts insulin sensitivity, so does chromium. Chromium derives its name from the Greek word for colour, and trace amounts of complexes of this metal are responsible for the colour of rubies; other forms are coloured intense green, yellow or purple. In the body, it works synergistically with insulin to facilitate blood glucose uptake. Such is its effect that insulin resistance is one of the symptoms that result from a deficiency of this nutrient. Food processing reduces dietary chromium and, because this mineral is lost in the urine, sweat, bile and hair, the body relies on regular doses from an external source.

Compared with chromium and vanadium, magnesium complexes are on the other end of the pretty spectrum and in the physical world shy away from contributing to colour. But in nature magnesium is part of the most abundant source of colour on the planet: it forms part of the green pigment in plants, known as chlorophyll. Unsurprisingly therefore, it is the seventh most common element in the Earth. Despite its abundance, dietary deficiencies are common and they have been linked to body weight gain. This micronutrient is important for making over 300 enzyme machines work correctly, including those involved in glucose metabolism, and its other roles include muscle contraction, nerve signalling and bone building, and so deficiencies have wide and varied symptoms stretching from muscle cramps to cardiovascular disease.[3] This latter effect is mediated via angiotensin II, a signaller that increases blood pressure. Magnesium deficiencies trigger angiotensin II overproduction, which in turn causes blood vessels to constrict, pushing blood pressure up. This signaller also triggers insulin resistance and weight gain.[4]

As with chromium, magnesium deficiencies are more likely to occur on a processed diet. Such diets generally lack rich sources, such as green vegetables, and processing removes many of the nutrients from the other foodstuffs, with, for example, the flour refinery process removing as much as 80% of the mineral content of grain. One-tenth of the people admitted to hospital have magnesium deficiencies.[5] Given that this number is just a fraction of the prevalence of obesity, these deficiencies are not alone in promoting unhealthy body weight, but may be a contributing factor.

Magnesium also regulates levels of calcium, which is the fifth most common element in the Earth's crust and the most abundant mineral in the human body. Magnesium works against this mineral in some regulatory functions. Calcium has been linked to the obesity epidemic, and supplements aid weight loss on a calorie-controlled diet.[6] It is also required for healthy bones, but bone-building is not its only important biological function; in fact, bones act as a reservoir to buffer calcium levels in the blood and cells, where it is a signaller. In the event of a calcium shortage the bone is stripped of this mineral because its signalling role is more important to survival than a decline in bone strength. Rickets is the disease that results from severe calcium deficiency and it is diagnosed by problems with bone formation and strength, but the changes in fat chemistry may occur silently before the signs in the bones. Calcium drives metabolism forwards and is a regulator of nutrient uptake into the body and then into cells. Thus, deficiencies of calcium have a knock-on effect on the availability of other nutrients. Excess dietary calcium, on the other hand, interferes with the absorption of magnesium from food, as well as causing a list of other debilitating symptoms.

Next to calcium, phosphorus is the second most abundant mineral in the body. It is the material of the elusive philosopher's stone; the slightly impure forms generated by alchemists glow with an eerie light and spontaneously burst into flames. One of the most reliable sources of the ingredients to make this mysterious substance was human urine. Phosphorus is there by no accident because this mineral is involved in filtering waste out of the body, as well as regulating the balance and storage of nutrients, including energy. Regular loss of this nutrient requires regular replenishment, and for much of human history phosphorus deficiencies were thought to be common. Yet modern diets contain foods and additives that are concentrated forms of phosphorous; rather than a specifically designed supplement, you can boost your daily dose by drinking colas. However, take this beverage with caution because phosphorus excesses are dangerous and some people are more sensitive than others. This mineral is needed in balance with calcium such that a cola too many can cause similar consequences to a calcium deficiency, including osteoporosis and, indirectly, weight gain; there is a certain irony when the word "diet" appears on cans of cola.

Zinc is another mineral required for the correct working of some enzymes, including those involved in protein, carbohydrate, fat and alcohol metabolism. It was probably named by Paracelsus, based on its close resemblance to tin, the German word for which is *Zinn*. Before it had this name, the metal was a fundamental part of many aspects of our ancestors' lives. It was mixed with copper to make bronze and in the body is essential for protein synthesis, the integrity of the cell membrane, DNA and RNA maintenance, tissue growth and repair, taste perception, prostaglandin production, bone mineralization, healthy thyroid function and blood clotting. Many of these functions could affect body weight regulation. The symptoms of severe zinc deficiency are wide and include glucose intolerance and skeletal abnormalities. Deficiencies of zinc often lead to poor appetite and consequent weight loss, but they also trigger a thrifty phenotype via changes in bone maintenance, glucose tolerance and thyroid function such that if the poor appetite can be overcome – perhaps by a fatty diet stimulating galanin – weight gain quickly follows.

In addition to being mixed together to make bronze, copper and zinc have associations in our bodies. They are absorbed via the same protein gates in the gut. The body can regulate copper intake by altering the number of gates, but when these two minerals are not balanced in the diet an excess of one can block the uptake of the other. The deficiency signals for an increased number of gates, feeding the excess – perhaps to a toxic level – whilst addressing the deficiency. The intertwined nature of nutrient uptake in the gut means that the body is unable to screen foods and is at the mercy of our mouths to promote the balance.

Copper is involved in processing the vitality nutrient iron and it is also needed to make collagen, the glue that holds our bodies together. Paradoxically, although inside the tissues copper and vitamin C work together in collagen maintenance, there is animosity inside the digestive tract: vitamin C interferes with the absorption of copper by promoting the formation of the poorly absorbed reduced form. Its interference with mineral absorption extends to calcium and vanadium. In this way excesses of vitamin C could interfere with mineral balance and body weight control.

Vitamin C is not the only vitamin that could affect body weight regulation; other members of this nutrient class have subtle, yet noticeable, effects. Chemically vitamins A, B, C, E and K are all related because they contain amines. Indeed, the common amine theme was the inspiration for their name. Towards the beginning of the 20th century, Kazimierz Funk, a Polish biochemist, coined the word vitamin from the term "vital amines", to describe a number of essential nutrients containing this chemical group. However, the honour of discovering vitamins was awarded to Eijkman and Hopkins by the Nobel committee in 1929 (see Left to our Own Devices: Third Time Lucky).[7]

Vitamin A is actually a group of chemically similar micronutrients that include the fat-soluble retinoids found in animal products as well as the precursor forms of the vitamin, also known as carotenoids. The carotenoids are the yellow–orange pigments that give certain fruits and vegetables colour. All forms of vitamin A have anti-oxidant properties and these nutrients have a popularised role in night vision. Excesses are poisonous.

Before the poisoning becomes lethal, it can perturb body weight and, like proteins, vitamin A seems to mediate its effect via thyroid function. In the blood, the thyroid hormone thyroxine (TY) circulates bound to one of three transport proteins: albumin, TY-binding globulin, and thyroid binding prealbumin, otherwise known as transthyretin (TBPA). Only when the thyroid hormones are released from the carrier proteins do they become active in the regulation of protein, fat and carbohydrate metabolism and how these fuels are used in cells. In this way, the amount of carrier protein can sway thyroid hormone function, and their carrying capacity is influenced by nutrition. In particular, dietary protein and vitamin A increase the amount of the circulating TBPA carrier. TBPA is usually responsible for transporting about 10% of the thyroid hormones, but if the amount of the carrier is raised, it could mop them up too efficiently and prevent their release when needed, dulling their effect and mimicking the consequences of an underactive thyroid gland. The altered amounts of protein in the system are likely to be transient adaptations in body chemistry in response to environmental signals.

The TBPA levels fluctuate in response to the amount of vitamin A in the diet because the carrier also transports retinol, a derivative of this vitamin. TBPA associates with retinol binding protein (RBP) for the transport of these nutrients around the body. Its role in vitamin A processing is critical, as demonstrated by the fact that mice completely lacking this carrier have just 5% of the normal levels of circulating RBP[8] and a compromised capacity to process vitamin A. When RBP levels are raised the fat chemistry profile has been linked to promoting obesity and type 2 diabetes.[9] The levels of RBP change in response to oral contraceptives and some types of infection, but mainly because of the amount of vitamin A circulating: they increase when needed to carry excesses around the body and decrease during vitamin A deficiencies. Thus, fluctuations of TBPA in response to dietary vitamin A appear to be an adaptive mechanism to allow correct processing of this nutrient – with a knock-on effect on body weight via the carrier's role in TY binding. Therefore, vitamin A can be added to the list of nutrients that in excess can trigger weight gain.

There are two sources of vitamin A in the diet: easily absorbed retinyl palmitate from animal fats and poorly bio-available carotenoids from plant foods. Because vitamin A is a fat soluble vitamin, it follows that foods containing this micronutrient also contain fat. Therefore, vitamin A-rich animal fats could be a double whammy against weight control, directly contributing to weight gain by providing calories and indirectly contributing via the effects of vitamin A on body chemistry. However, there is another possible explanation for the link: circumstance. Excess fat causes weight gain and the wide use of fortification suggests that in that excess fat there is likely to be excess vitamin A boosting RBP levels. Rather than raised RBP "causing" obesity, it may just be another symptom of the poor nutrition that leads to excessive weight gain, and to insulin resistance. However, dietary regulation of this nutrient is important: too much can poison the body, with mild excesses causing problems such as dry lips, hair loss, easily broken nails, bone pain, gingivitis and liver problems and extreme excesses resulting in death. Like all nutrients, the benefits of vitamin A are dose dependent.

The B complex is also a collection of vitamins grouped together, partly because they are found in the same foods: whole-grain cereals, meat, eggs and so on, the foundation foods in a balanced diet. The group includes pantothenic acid, niacin, biotin, folic acid, riboflavin, thiamine and cobalamin; all complex names to describe vital amines with wide functions including promoting healthy skin and muscle tone, immune function, nervous system health, and cell growth, including production of red blood cells to prevent anaemia. There is also a link between the B vitamins and body weight: members of the family are involved in metabolic regulation and mood, such that imbalances could affect appetite and, thereby, body weight. Niacin and riboflavin are involved in harvesting the energy released from food and transforming it into a form more readily accessible to the cells as the citric acid cycle turns. These micronutrients also form part of the molecules that store the energy: NAD and the closely related flavine adenine di-nucleotide (FAD), respectively. Thiamine acts as a co-enzyme to keep the cycle turning, and pantothenic acid (vitamin B_5) forms part of the co-enzyme A structure.

Vitamin C is a water-soluble anti-oxidant responsible for protecting the body against toxins and for recycling iron and vitamin E, extending their usefulness. Vitamin C is also involved in the production of bile from cholesterol. Deficiencies interfere with these functions causing scurvy, which in extreme cases results in the breakdown of the collagen glue that holds the body together, one of the first symptoms of which is fragile veins. Before bruising and haemorrhage occur, there are specific consequences of inadequate doses in the diet, one of which is compromised immune function. Vitamin C inactivates toxins known as oxidants – it functions as an anti-oxidant. Without this defence against chemical assaults a whole range of degenerative diseases, from heart disease to ageing, are more likely to occur (see A Question of Dose: Chemical Assaults and the Clean-up Process).

Given its anti-oxidant function alone, vitamin C is hyped as a super-supplement. Its water-soluble nature means that excess is flushed out of the body; hence, it is also hailed as a supplement you cannot take in excess. Promotions of the health-boosting properties of vitamin C have been so successful that it is now available in mega doses. The current recommended upper limit for the safe intake of vitamin C is about 1000 milligrams per day. Supplements are available in 1000 milligram doses, which leaves no quota for dietary sources: no vegetables, no fruit, no potatoes and no fortified foods. Realistically, no one has a diet devoid of vitamin C and, therefore, supplements that contain the full daily recommended dose of this vitamin potentially lead to overdoses.

Nutrient excesses can be as harmful as deficiencies, as Paracelsus observed several hundred years ago. The popular view is that the water-soluble nature of vitamin C allows the body to flush out what it does not need, to reduce the effects of poisoning. However, the analysis focuses on the vitamin itself and not the downstream effects on other nutrient levels nor their effects on fat chemistry. Excess of this nutrient has been linked to stomach cramps, nausea, kidney stones, perhaps even leukoderma (where the skin starts to peel off) and obesity. Many of these effects are a consequence of the flushing process removing other nutrients, particularly those whose uptake depends on their oxidation state; vitamin C not only performs its anti-oxidant activity against harmful chemicals, but also minerals. The obese require more vitamin C than those in the healthy BMI zone because of altered gene use. This increased requirement needs to be matched with a higher dose of minerals, including copper, calcium and magnesium. In this way, the altered nutritional requirement caused by increased body weight makes heavy folk more susceptible to the types of malnutrition that promote body weight gain, trapping them in obesity.

Vitamin E is vitamin C's partner in defence: it is another anti-oxidant, but it is fat-soluble. It has a key role in maintaining the integrity of fatty parts of the body, including cell membranes. It enhances immune function by preserving the cells involved in defence and it protects against premature ageing, cataracts, diabetes, cardiovascular disease, inflammation and infection. For example, the vitamin E content of low-density lipoprotein (LDL; see The Image of Fat: Body Fat) protects the cholesterol component from oxidation, thereby inhibiting the initiation of plaque formation in the arteries. Because it is found in fats, low-fat diets can lead to deficiencies of vitamin E. Additionally, there is a fat quality factor: it can be destroyed by food processing such that low quality high-fat diets can still lead to deficiencies. Even if ample vitamin E enters the gut, absorption problems can cause deficiencies: disease or drugs that interfere with fat absorption block its uptake, as do certain micronutrient imbalances. The absorption of vitamin E also depends on correct production of the bile acids, which in turn depends on iron and vitamin C levels. Yet despite its health-boosting effects, excesses have been linked to obesity: like vitamin A, vitamin E pushes up RBP levels. Unlike the water-soluble vitamins, the hydrophobic counterparts are not easily excreted. They accumulate in the body. Their poisoning effects can be reduced by locking them up – along with other fat-soluble toxins – in fat deposits. This sequestering function is pro-portional to the size of the store, with larger deposits locking up more nutrition. These chemicals are released as the fat stores melt away. Therefore, if vitamin E promotes obesity, as lipids are resourced its release may impede further fat metabolism, main-taining adipose stores at a constant level.

The last vitamin is vitamin K, a fat-soluble amine first identified as essential for blood clotting. The alphabetical designation of this vitamin is derived from the Danish word for coagulation, because of its function, rather than the order of its discovery. It also acts synergistically with vitamins A and D in bone main-tenance. The microbes in the gut provide most of the vitamin K the body needs; hence, long-term antibiotic therapy, chronic diarrhoea or impaired gallbladder function can reduce uptake, but because this vitamin is storable – fat soluble – debilitating deficiencies are rare. The classic symptom of deficiency is slow blood clotting, increasing the chance of haemorrhage, but the symptoms may not be due to an absolute deficiency but a relative one: vitamin E excesses can inhibit vitamin K activity. Addressing this imbalance is tricky: because of its role in clotting, vitamin K supplements are too risky to prescribe to the general public as excess can lead to thrombogeneisis.

Despite its name, vitamin D is not a true vitamin because it is chemically different: it is not an amine. Amounts of this nutrient are obtained directly from food, naturally and because of fortification, but most of the body's needs are met by in-house production, harnessing the energy in sunlight to make vitamin D_3 from 7-dehydrocholesterol, a familiar sounding name, part of which is "cholesterol". This name did not occur by chance, because one is a derivative of the other. The body makes up to 1 gram of cholesterol daily – more than most people eat – which accumulates therein to about 150 grams, a quarter of which is lodged in the membranes that surround cells regulating their function, and the rest is distributed around the body for the synthesis of sex hormones, bile acids and vitamin D.

Vitamin D is most famed for its role in bone maintenance. However, its contribution to good health is diverse. For example, it stops potential cancer cells in their tracks. In fact, it stops all cells from dividing. Cell division is an essential step in the growth process; cells grow in size until they reach the maximum for efficient feeding and waste disposal. At this point they split to form two half-sized cells that enlarge until the full size is reached. Vitamin D stops the cells from splitting; hence, it stops cells increasing in number.

Over the past 20 years or so, since vitamin D was first shown in the laboratory to slow the growth of leukaemia cells, it has also been shown to have a similar affect on colon, prostate, breast, lung and skin cancer cells. In a 19-year epidemiological study, the colorectal cancer rate was shown to be halved in the 1000 Chicago men with the highest dietary vitamin D intake.[10] So important is the effect of this food component in cancer prevention that pharmaceutical companies are developing vitamin D-like anti-cancer drugs.[11]

The cancer-stopping effect of vitamin D is not strictly essential for our health; it is just an added bonus. In the same way, B vitamins have bonus value – promoting health beyond nutrition – by reducing the symptoms of stress, depression and cardiovascular disease. These properties are not essential for the functioning of the body, at least not over the same timescale as promoting red blood cell growth to prevent anaemia, but for some people, the stress-busting affects of this vitamin are important. There are many other food components that have bonus value without being essential nutrients and these components are often termed functional foods. Functional foods are not needed by the body, but they can improve health. The group includes all the latest anti-oxidants and phyto-chemicals classed as beneficial to good health, but not absolutely necessary. Their functions may be interchangeable with those of other nutrients; the benefit to health may be slow, acting over decades rather than years; or their effects may be more cosmetic than vital. That some food components benefit health but are not essential and other foods are essential yet the deficiencies only manifest themselves over long timescales is an important point when quantifying nutritional needs: do we include nutrients that give a health boost, a long-term survival advantage, or stick to those that are essential in the short term, such as those Hopkins and Eijkman are famed for finding?

Functional foods include phytochemicals, which are relatively complex chemicals usually made by plants, although the classifi-cation is not exclusive because some animals can make some phytochemicals too. Plants and animals make these nutrients for their own purposes: some are needed for metabolic regulation, others for signalling and coordination, and others are natural pesticides. Many of these chemicals have health benefits; for example, some phytochemicals called flavonoids have been shown to reduce the risk of developing cancer.[12] The natural production of phytoestrogens – the phytochemicals that mimic human hor-mones – protects plants from stress. In the human body they can help to protect against osteoporosis, cancer and heart disease. The phytoestrogens are not essential for human health, but their well-being boost could make them a useful addition to the diet. Bio-flavonoids are anti-oxidant phytochemicals produced by plants to protect themselves from the damaging effects of the vital gas oxygen. They have similar protective benefits in humans. They include coenzyme Q10, one of the trendy additives to skin and hair-care products. Coenzyme Q10 is made in the body as it is needed to release energy from food, but because it can be made in-house it is not classed as a nutrient. Yet its properties suggest that a little more may boost health: this chemical is a fat-soluble anti-oxidant (see A Question of Dose: Chemical Assaults and the Clean-up Process). Hence, a little more in the diet may improve health, but too much may be detrimental, perhaps altering nutrient uptake in the same way as vitamin C excesses.

Phytochemicals in broccoli kill the bacteria that cause stomach ulcers; the broccoli plant makes these compounds to protect itself against infection and they still work when the vegetable is cooked, mashed and enters the digestive system. These phytochemicals are not essential for good health, but their addition to the diet may protect the body against disease. Then there are groups of phytochemicals produced by cruciferous vegetables as insecticides that also offer cancer protection. Usually these chemicals are harmless and the dose is regulated by their bitter taste. In the body, they promote the development of protective mechanisms against harmful chemicals, both natural and synthetic, that could cause cancer. The effect comes down to the fact that eating these natural chemicals changes the way the body uses its genes. The pathways that process the bitter tasting phytochemicals in Brussels sprouts can also deactivate carcinogens. Without vegetables in the diet, these deactivating pathways are an extravagance and are rarely used; for this reason, they are not maintained and the capacity to prevent damage caused by carcinogens falls too. So significant is their effect that there is a statistically higher chance of developing cancer if the diet lacks these foods.

A lack of vegetables in the diet leads to a reduction in the efficiency of detoxification of poisons. Without rapid neutralisation, harmful chemicals hang around, do a bit of damage, maybe trigger a chain reaction that causes even more damage, and then do a bit more damage before the body can assemble what it needs for deactivation. Each bit of damage raises the chance of developing cancer and promotes ageing and other chronic diseases. In contrast, people who eat many vegetables have these processing pathways up and running to deal with the harmless, bitter tasting phytochemicals as they come along, preparing their bodies to deal with all sorts of chemical threats. Eating vegetables promotes investment in a rapid response team – even a super hero – waiting for any toxin that should pass by to swot it, arrest it and push it out of the body as quickly as possible before it causes any harm. These bitter tasting phytochemicals are not essential for the body's functioning in the same way as calcium and vitamin C, for example, but they offer a valuable health benefit.

Some food components have beneficial effects related to unhealthy body weight and its associated health problems, directly targeting metabolic and regulatory pathways. For example, the seeds of the *Argania spinosa* tree have insulin-sensitising properties. These thorny twists produce slow-ripening, thumb-sized fruits with a thick, bitter peel surrounding a sweet-smelling but foul flavoured pulp. Get past the peel and the pulp, and you will find a very hard nut which is the treasure of these trees. Several women's co-operatives in southwest Morocco specialise in cracking these nuts and extracting the small, oil-rich seeds to roast and press. The oil has been used in traditional Moroccan medicine for centuries to treat diabetes and now is marketed as a health product worldwide.

More accessible insulin-sensitising effects can be given by choosing traditionally spiced foods. Cinnamon was used to amplify the sweetness of foods when sugar was rare. However, the complementary effects of cinnamon and sugar go further than taste: phytochemicals in cinnamon lower blood sugar, reactivating broken insulin receptors in the diabetic's body. Richard Anderson of the Human Nutrition Research Centre in Maryland, USA, stumbled upon this discovery by accident. He was studying the effects of common foods on blood sugar levels, one of which was traditional American apple pie. The studies involved feeding volunteers their favourite foods in exchange for a blood glucose test after the meal. Despite the high sugar content of the pie, the blood sugar levels of the volunteers did not shoot up as expected. Further studies showed that cinnamon spice – not the artificial flavour equivalent – contained a phytochemical that worked alongside insulin to enhance its effect to such an extent that some type 2 diabetics no longer needed drug therapy (although they did have to eat a great deal of cinnamon). However, Anderson was not the first person to discover the medicinal properties of this spice: in Sri Lanka cups carved from cinnamon wood are a traditional wedding gift and daily drinks prepared in them are said to prolong life, possibly by keeping diabetes at bay.

In addition to lowering blood sugar levels, cinnamon lowers the levels of fats and the amount of cholesterol shipped out from the liver in LDL (see The Image of Fat: Loops and Rings). The relative amounts of these fats are partly controlled by insulin and can affect appetite via the release of this hormone. This spice also has some anti-oxidant effects. However, before rushing out to stock up on cinnamon, remember that research into just about every natural food would give some promising results for improved health, because, after all, these are the foods on which the body is designed to thrive.

Some soy phytochemicals affect energy metabolism,[13] and others are thought to be partially responsible for the cholesterol-lowering and the anti-diabetic effects of soy in the diet.[14] Soybean phyto-chemicals also mildly block starch digestion, reducing the amount of glucose released from the food and thereby reducing the amount of energy available to the body.[15] Soybean sprouts are the most efficient in this respect, possibly because these phytochemicals are most important to the plant at this stage in their lives. Indeed, these phytochemicals may be part of the plant's protection mechanism, designed to do the average animal that chooses to eat bean sprouts no favours by reducing the amount of valuable energy available from the meal, thereby reducing survival when food is in short supply. Of course, the impact of bean-sprout eating depends on the animal's size and relative food scarcity, but these phytochemicals may protect uneaten sprouts by killing off their predators, increasing the number of sprouting seeds available to produce the next generation of plants. For people, however, where food is plentiful and energy needs to be restricted, this defence mechanism may actually be doing us a favour helping to curb weight gain; although given our size, the digestive effects of a normal portion of bean sprouts is likely to be hard to detect.

Phytochemicals also have beauty benefits, such as those due to alpha-lipoic acid, a naturally occurring, modified fatty acid said to reduce wrinkles in the skin. Our bodies can synthesise this fat, so it is not an essential nutrient; indeed there are many questions as to whether the body can absorb it from food at all. Still, it is available in supplement form for a wide range of complaints, including weight loss, and as an energy boost. Rutin, found in asparagus and citrus fruit peel, for example, is proposed to strengthen capillaries and reduce broken veins; a cosmetic effect. There is some evidence that it also reduces angiogenesis, an important stage in cancer development. These are two examples where classification is unclear. Are these nutrients "beneficial", "valuable" or "vital"? Most people would say that the right balance of alpha-lipoic acid and rutin is beneficial, but not essential, because their effects are mainly cosmetic. But, how essential are the phytochemicals in soy for lowering cholesterol or those in cinnamon for offering protection against diabetes? Some people would say that they were valuable, but not essential because they only help to overcome insulin resistance, which in turn may be a result of other inadequacies in the diet, for example vanadium deficiency. However, many diabetics who control their insulin resistance with large doses of cinnamon may disagree. Thus, differentiating between nutrients that are beneficial, valuable and vital for good health is complicated by individual needs and perspectives.

The micronutrients we need are found naturally in pretty much the same foods as the components that offer a health benefit that is classified beyond our needs. Our bodies are apt at extracting what they need from this sort of food. Therefore, a holistic diet negates the need to dwell on details. The same truth does not apply to a more artificial situation where conscience-clearing supplements are used to justify a diet that contains mainly nutrient-poor foods. In these situations defining what we need becomes of critical importance – and an enormous responsibility. What do we need? Our bodies need many nutrients to stay healthy and the level of health and longevity may be improved by adding functional foods to the menu. But the need is not just qualitative; the quantity of nutrients is important too. After all: Everything is poisonous, and nothing is harmless. The dose alone defines whether something isn't poison – Paracelsus (1493–1541).

7.2   GENERALLY RECOGNISED AS SAFE

Paracelsus' theory of dose and effect goes beyond the beneficial nutrients to include all food components. Over the past century new ingredients have been introduced into the diet that are neither micronutrients nor functional foods. These ingredients offer other benefits: increased shelf life, appealing colours and textures, or a divine flavour. Before becoming part of a food or beverage, each additive is assessed for potential health effects. Usually the results of this assessment are vague: the additives are assigned the categorisation of "generally recognised as safe" (GRAS), which means for the intended use an additive has no known detrimental safety issues; the dose is below that of a poison. The protection such a categorisation offers depends on the definition of safe. No one would hesitate at evicting from the GRAS list a chemical additive that quickly caused cancer, diabetes or heart disease via its intended use (even though some natural food components – saturated fat, cholesterol and sugar, as examples – may have the same effect). But what about the subtleties: chemicals that alter flavour perception and interfere with our inbuilt protection against malnutrition, or chemicals that contribute to multi-factorial illnesses, such as depression or body weight gain? These effects may take a few decades to be noticeable.

The GRAS categorisation also assumes that the food products are incorporated into a balanced diet. When that balance no longer exists, problems can arise both because of the amounts of individual components and because of the cumulative effects of several additives. And then there is the "cocktail effect". The cocktail effect describes what happens when two or more food components are mixed. In some cases nothing happens: although the effects of additives combine, they do not chemically react with each other, making the risk assessment easier. However, in some cases the individual food components react together to produce new chemicals. These reactions are behind the synergistic flavours of food or drink combinations, and give cocktails a distinctive character when they are shaken rather than stirred. But changing the nature of the chemicals in foods may introduce health-promoting or health-damaging components.

The body weight boosting effects of any one additive are often negligible – they can be generally recognised as safe – but there are examples where these effects are cumulative with those of other additives; cumulative to such an extent that they may override appetite control. Take, for example, the interplay between phosphorus, calcium, vitamin D, glutamate, aspartate and salt. Phosphorus is an abundant mineral in the body, but most is locked up in bone. It is usually found bound to oxygen to form phosphates. Free phosphates are produced naturally in the body, for example when ATP is cleaved to release trapped energy. The amount of the free mineral is important for regulating protein function and calcium balance: an excess signals bone weakening,[16] releasing calcium for its signalling function and thereby altering body chemistry. Via bone strength the body gauges its nutritional status, altering appetite and metabolism. As bones begin to weaken, the energy balance becomes biased in favour of weight gain. Phosphorus is concentrated as a food additive and is a principal ingredient in cola; for many children cola has replaced milk as a snack time refreshment. A main manufacturer of cola states that, in a standard can, there is one-fifth of the daily recommended intake of phosphorus for a sensitive adult; for the rest of us, we would need to drink 17 cans to get the full daily quota. However, a supersized drink (about 1.25 litres of liquid) approaches the tolerance limit for a sensitive person (and is also close to the calorie content of a meal). Few people with common sense – sensitive or otherwise – would try to drink a supersized cola alone; and yet these products are marketed as an individual serving, suggesting that they are safe. The healthiest dose of phosphorus depends on many factors: genes can affect sensitivity and so can diet, such that a single standard can of cola may not have a detrimental effect on health, but combined with other phosphorus-containing food additives and a low calcium intake, the effects become more substantial.

Glutamate is an amino acid building block of proteins and a signaller. It is a neurotransmitter involved in learning and memory and it regulates calcium signalling in the brain, but it is also used by bones to regulate their growth. Glutamate-signalling is stimulated in response to calcium levels and mechanical loading on the bones, to communicate the needs for remodelling;[17] it is one of the signallers that ensure that bones obey Wolff's law. It is also a component of the widely and generously added flavour enhancer monosodium glutamate (MSG). During cooking, protein breaks down and this building block is liberated in its free, flavour-enhancing form. MSG is, therefore, based around a natural food component, but concentrated to give a large dose; just like dietary supplements – and poisons. Alone, a Chinese meal laced with soy sauce – one of the most concentrated sources of MSG – may not have a noticeable effect on health, but combine it with a can of cola on the background of a calcium-deficient diet and the effects on bone strength begin to accumulate, and so do the knock-on effects on calcium signalling and body weight regulation.

Glutamate is not the only signalling amino acid that is concentrated as a food additive. Aspartate is another example. Whereas glutamate is a savoury flavour enhancer, aspartate is in the sweet camp; but rather than being useful in the free form, it is modified to make the artificial sweetener aspartame. Despite the flavour differences, the two amino acids are similar enough to interchange in protein structures without, in the most part, catastrophic consequences; they may be similar enough for aspartate to mimic glutamate's calcium-stripping effect in the bones. Certainly both amino acids can perform some of the same signalling functions in the body, including memory-making.

Memory and learning involve the glutamate N-methyl-D-aspartate (NMDA) receptors in the brain. These receptors are involved with recording the coincidence that hunger is satisfied when we stand in front of a hot-dog stand when it is open but not when it is closed. Aspartate and, more powerfully, glutamate activate NMDA receptors in the brain. Injections of glutamate have been shown to enhance some types of memory formation in rats,[18] but whether the food-form helps us to remember the tasty morsels in which it hides is difficult to say. In mice, the memories involving the glutamate pathway include those of the smell of male pheromones during mating,[19] intertwining food and love. However, very high doses of either glutamate or aspartate will impair memory function and have even been linked to neuron damage;[20] a poisoning effect. In any single food product, the amount of aspartate is GRAS, but when lunches are themed around cola drinks with artificial sweeteners and soy-sauce noodles the amounts could exceed individual tolerance levels, altering memory and eventually poisoning these same signalling pathways.

With the obesity epidemic there has been an increase in the number of cases of learning problems such as attention deficit hyperactivity disorder (ADHD): in the USA the number of cases of ADHD increased by 2.5 times between 1990 and 1995.[21] Part of the increase could be attributed to diagnosis, but there are other suggestions: dietary MSG is one; artificial colours and flavours, chocolate, preservatives, caffeine, simple sugars and lactose are others.[22] Some of these food components are also linked to promoting body weight gain and, therefore, the concurrent rise of these two issues – obesity and ADHD – may be coincidence or it may be a consequence of the same cause.

In addition to the accumulating doses of phosphorus, aspartate and glutamate altering bone strength, appetite and body weight, sodium, half of humble table salt, also affects the NMDA signalling cumulatively with these nutrients. Sodium, balanced with potassium, is involved in the clean-up process during glutamate signalling. Excess sodium mimics and accumulates with the effects of excess MSG. We are genetically programmed to enjoy this seasoning (and become accustomed to the flavour), because sodium is an essential nutrient rare in the natural world. Salt is so cheap there is no restriction on the amount we can add to our food – we can eat as much as we like! But the dose chosen by much of the world's population is health damaging: salt is the single additive that the World Health Organization (WHO) chose to target for reduction campaigns in an attempt to improve human health. This is an important point: when experts studied nutrition to identify the single most effective target to improve global health they realised it was not the chemical additives so often blamed in the media – the ones that the food industry actively tries to estrange – but salt. The message that one of our Stone Age ancestors' favourite flavours promotes heart disease is beginning to filter through some 24 centuries after the Chinese had noted that a high salt diet "hardens the pulse";[23] let's hope the new awareness campaign begun by WHO does not take so long to improve world health.

Salt, glutamate, aspartate and phosphorus are essential nutrients that form part of additives that are generally recognised as safe: they are additives based on natural and essential components of food. But their cumulative effects suggest limits to GRAS protection. Safety depends on the dose; not just individually, but the relative dose in balance with the availability of other nutrients.

Aspartate-containing sweeteners, phosphates, MSG and salt are added to many foods that have experienced manufacturing. Individually they are added in amounts that are safe. But the cumulative effects of these nutrients on health are becoming increasingly difficult to gauge, particularly with respect to their effect on the balance of nutrients known to be needed for good health. Phosphate and glutamate interfere with calcium and vitamin D balance and deficiencies of these latter two micronutrients have been linked to unhealthily high body weight – not just in causing the problem, but in perpetuating the disease. Recall that the demand for vitamin C is increased among the obese because of the increased need for collagen repair (see Beyond the Helix: The Power of Epigenetics). Similarly, nutritional needs with respect to vitamin D alter as fat deposits grow. The availability of vitamin D is decreased among the obese because this fat-soluble nutrient is locked away proportionally in fat.[24] The knock-on effect of this altered fat chemistry causes falling calcium levels because vitamin D is needed for the absorption of this mineral in the gut. Together, vitamin D and calcium strengthen bones and the body gauges nutritional status via bone strength. However, calcium is also involved in charging ATP molecules with energy by forming gradients across the mitochondrial membrane that allow ADP to capture a third phosphate group and form ATP. Calcium is also a signaller. It is via this signalling that phosphorus (most often in the form of cola), glutamate (usually in the form of MSG) and sodium (from salt) have a cumulative effect with aspartate (in the form of the artificial sweetener aspartame) to promote body weight gain. On this line of thought, it is also worth mentioning that several forms of obesity treatment, including biliopancreatic diversion and Roux-en-Y bypass, aggravate these imbalances.[25]

Authorities who evaluate food safety consider the health consequences of all chemicals that come into contact with our bodies. With the rise in the incidence of obesity, there has been a technological revolution that has exposed humans to a variety of new chemical insults – exhaust fumes, cleaning products, chemical residues in foods and so on – and as a result, the nutritional requirements of the human race may have changed. Bisphenol A (BPA) is a chemical used in the manufacture of polycarbonate plastics that received much media attention because it leaches into food stored in these products. Once there it is thought to cause health problems. For example, perinatal exposure has been linked to a higher adult body weight, an increased risk of developing breast and prostate cancer, and altered reproductive function.[26] It is important to keep these frightening findings in perspective: much of the human race has received this exposure, but only some have become obese and even fewer have developed cancer. Some of Jirtle's studies suggest that the difference in susceptibility may be due to nutrition.[27] Among the agouti mice, BPA exposure promotes a yellower, fatter profile, but nutritional supplements completely nullify this effect: nutrition protects against the obesity-promoting effects of environmental chemicals. Although the *AgRP* gene is unusual, the findings are interesting, not least because dietary-induced changes in appetite and metabolic regulatory proteins could explain the epidemiology of unhealthy body fat. The profile of the current obesity epidemic in industrialised nations suggests the cause is not global and passive, for example the weather or pollution, because obesity may affect one man, but not his neighbour. Rather, obesity seems to be triggered by personal factors, such as diet and lifestyle, and clusters amongst families could be as much due to genes as traditions; the gene-pool was largely the same before obesity became epidemic, but chemical exposure, lifestyle, nutritional needs and diet have changed.

7.3   CHEMICAL ASSAULTS AND THE CLEAN-UP PROCESS

Dietary requirements are individual and they change according to lifestyle and health; they even change according to body weight. The direction of the change is to promote certain deficiencies that are thought to perpetuate obesity. Because of changes in fat chemistry, the obese have a higher requirement for vitamins C and D than their trim neighbours, and both vitamins can be linked to weight regulation. For monkeys, when their mums are fed on a high-fat diet there are a number of significant changes in metabolism, including a reduced circulation of vitamin C,[28] which could be explained by MMP9 gene over-expression increasing the requirement for this nutrient in collagen synthesis. In addition to collagen repair, the vitamin C and E requirements are generally raised among the obese because obese bodies burn more fuel than if they were a healthy weight, creating more oxidants that need to be deactivated.

Oxidation is the chemical reaction that causes oxidative damage. It involves the exchange of those tiny atomic particles called electrons. The name of the process is derived from the element oxygen because the most familiar oxidation process is burning – combustion – and oxygen is the oxidant in that process.

Recall that atoms, including those of oxygen, have a nucleus at the centre that contains most of their mass, surrounded by a considerably lighter and diffuse cloud of electrons (see Why the Fuss About Obesity? Clearing the Name of Chemistry). The number of protons determines the atom type and, in a theoretical single atom, the number of electrons and protons match so that there is no overall charge. The architecture of the atom means that the electron orbitals are ordered in shells. The first shell houses two electrons; the second shell can house up to eight. Therefore, if an oxygen atom has eight electrons, they will sequentially fill these shells, two in the first and six in the second. But electrons like to be paired. For an oxygen atom, four of the six electrons in the second shell will form two pairs and the other two will be alone in an orbital each and looking for liaisons. The architecture does not allow the singles to pair with each other. Therefore atoms of oxygen make liaisons with other atoms with similarly single electrons, perhaps a second oxygen atom, locking together the race-track orbitals so that the single electrons can be paired and shared between the atoms.

A molecule of oxygen, as it is found in the air that we breathe, is composed of two oxygen atoms, each sharing two electrons with their partner. Water molecules are formed by an oxygen atom sharing electrons with two hydrogen atoms. But there are other ways of satisfying loneliness. Oxygen can pinch electrons from other atoms to fill its race-track orbitals, leaving the other atom with completely empty tracks. When it does steal outright, it has extra electrons compared with the number of protons in the nucleus and so carries a charge. This charged species is called an ion. Ions are ubiquitous in nature and are often responsible for colours, from the luminescence of the Sun to the colour of gem-stones. Stable forms of these charged species can be made from some atom types, but not all. Sodium chloride, table salt, is made from positively charged sodium ions paired with negatively charged chloride ions and packed in a crystalline structure held together by their opposite charges. Sodium is typical of metals, forming positively charged ions, and chlorine is typical of non-metals, forming negatively charged species. Oxygen is a typical non-metal.

In addition to sharing two electrons in a friendly way and pinching electrons to form stable ions, oxygen can form some less stable species. In peroxides, each atom in molecular oxygen has managed to acquire an additional electron; it is no longer a stable molecule but a reactive species. Superoxides are even more capricious, because the oxygen atoms in this faction have not even managed to gain a whole electron for themselves; instead, they share a stolen electron across the two atoms of an oxygen molecule. The peroxides and superoxides are on the lookout for electrons. Superoxides are so dedicated in their search they are called free radicals and peroxides can be easily modified to join this band. The problem with free radicals is that once formed they are not easily destroyed without the right defences. If a free radical pinches an electron from another species, perhaps a protein or a piece of DNA, it pairs its own free electron, but now the victim of the theft is itself an electron short and may even be a free radical. In turn, the victim looks to steal from another molecule, creating a chain of destruction because, more often than not, passing through a free-radical stage destroys the molecule's correct function.

To biological systems, superoxide is toxic, but its toxicity can be controlled. A central part of the immune system involves generating superoxide to kill invading microorganisms. When specialised white blood cells called phagocytes engulf pathogens, they are isolated in a special bubble where an enzyme called an oxidase pinches electrons from the co-enzyme nicotinamide adenine dinucleotide phosphate (NADPH), a molecule similar to NAD, and then gives its bounty to the molecular oxygen dissolved in the bubble, to create superoxide. The flood of electron-searching free radicals chemically burns the bugs; its toxicity is channelled to our advantage, and oxidation – the burning process – is controlled so that it does not destroy the organism, only the pathogen. One of the principal free-radical scavengers, superoxide dismutase (SOD) enzymes, are on the ready outside the burning bubble to deactivate any free radicals should they leak out. These enzymes are also involved in the day to day running of the cell because anywhere there is oxygen in a reacting pot of chemistry – as there is in all our cells – there is the potential to generate superoxide and this powerful oxidant needs to be efficiently deactivated before it begins to burn the cell from the inside out. The turning of the citric acid cycle – when molecular oxygen is sequentially added to the carbon atoms in food to allow the energy trapped therein to be transferred to more useful forms – generates superoxide. Inside the mitochondria, where this process takes place, free-radical scavengers are stashed to neutralise the dangerous chemicals almost as instantly as these two species collide. Without this neutralising activity, the prognosis is poor: mice lacking SOD in their mitochondria die when they are just three weeks old as a result of neurodegeneration, cardiomyopathy and lactic acidosis.[29] When the deficiency of the enzyme is localised in the main body of the cell, the mice last a little longer, but develop multiple pathologies, including liver cancer, muscle atrophy, cataracts, haemolytic anaemia and a very rapid age-dependent decline in female fertility.[29] In fact, all the symptoms can be attributed to premature ageing; thus, SODs are enzymes that protect against degeneration and ageing. Anti-oxidants are postulated to give them a helping hand.

Anti-oxidants are now a household name, used in beauty products and food allegedly to protect our bodies against the harmful damage caused by oxidation. Vitamin C, otherwise known as ascorbic acid, is a common example. This anti-oxidant absorbs the damaging potential of oxidants by electron transfer reactions – electron juggling. The trick allows this micronutrient to donate an electron to another molecule without itself becoming a free radical. Therefore, when superoxides and free radicals are out to pinch electrons from wherever they can, vitamin C gladly provides and in doing so quenches their damaging potential. A single electron is released to satisfy the free radical and the loss is buffered by electron transfer reactions around what was vitamin C and, an electron down, is now called semidehydroascorbate. Subsequently, this species donates its single electron to neutralise a second free radical before being shipped to the endoplasmic reticulum (ER) compartment via transporters that usually transport glucose. Inside the ER, a multifunctional molecular redox molecule, named glutathione, donates electrons to reform vitamin C, which is then returned to the mitochondria for more radical action. Vitamin C works alongside enzymes such as SOD to scavenge free radicals, thereby reducing their potential to damage proteins and DNA. Like other anti-oxidants it is concentrated in the mitochondria. When the level of protection fails to meet the demand the damage that results has been linked to the development of metabolic diseases, including obesity. One of the most vulnerable groups of people to fall foul to the effects of oxidation are the obese, because of both their greater requirement for vitamin C for collagen repair and their raised level of fuel consumption that statistically increases the requirement for anti-oxidants. In this way, there is a cyclical mechanism that steadily promotes body weight gain.

The mechanism by which oxidation mediates its obesity-promoting effects has long-term consequences on health for the obese and their children, because oxidation not only mediates its effect via direct damage to the biomolecules of the cell, but also via in changes to the epigenome. The chemical memories are not annotated directly on the DNA molecule, but on the histone proteins around which it wraps to protect it from snapping. The information storage device is coiled around and around histone bobbins, kept tightly packed and held in place by a charge attraction: DNA is negatively charged and histone proteins are laced with positive equivalents. Protein charge is a property of its amino acid building blocks: glutamate and aspartate amino acids usually carry a negative charge and arginine, lysine and histidine a positive charge in the environment of the cell. Histones are made with a large proportion of positively charged amino acids, particularly histidine, hence the name. As with many substances in biology, the roles of histones are numerous: going beyond packing DNA to include regulatory functions. Methylation is one mechanism of regulation. Tagging these chemical blocks onto positively charged lysine or arginine amino acid residues in the histones neutralises the positive charge, relaxing the structure.[30] This tag alters gene use and the pattern of methylation can be inherited.[31]

There are other chemical modifications to histones that regulate gene use. Remember acetyl-co-enzyme A? In addition to fuelling the citric acid cycle and fat synthesis, acetyl-co-enzyme A can alter the way the cell uses genes. It regulates genes by tethering its acetyl units to lysine amino acid blocks on the histone protein chain. Histones regulate gene use by the tightness of their grip on the DNA strand: the more tightly wrapped they are, the less readily the DNA reading machinery can do its job. The acetyl units carry a negative charge, so they neutralise the positive charge of the lysine and thereby reduce the charge of the histone protein, decreasing its affinity for negatively charged DNA. The structure relaxes, increasing the accessibility of the information thread to the DNA-reading machinery. In this way, acetylation increases gene use and the permanence of the measure means that it is categorised under epigenetics: histone acetylation patterns are certainly stable enough to persist as the cell divides.[32]

Increased gene use via histone acetylation seems to lead to obesity.[33] At first glance, this link may be considered to be a coincidence because the rate of both histone acetylation and fat synthesis is dependent on the same cofactor, but there is evidence that suggests it is not just a concurrence. First, there are parallels that can be drawn with other studies. For example, Jirtle demonstrated that, for the Agouti mouse, elevated levels of gene use via changes in the degree of methylation on the DNA itself correlated with a raised chance of becoming obese. These nutrients mediate their effect with little calorific value. The relationship between gene use and obesity is the same irrespective of whether the cause is virtually calorie-free vitamins or cellular fuel: increased gene use increases body weight. Second, there is molecular evidence to suggest that the effects are due to the altered gene use and not the calorie supply. Just as the raised sirtuin levels caused by calorie restriction lead to epigenetic changes that promote weight loss (see Stone Age Obesity: Obesity Resistance), epigenetic changes amongst the obese involve the *sirtuin* gene. And the relationship involves nutrition, but not calories.

Sirtuin has several functions including histone deacetylation. When animals inherit polymorphisms of this gene that result in its protein's function being blocked, they lose their capacity to remove acetyl groups tagged on the histones, raising gene use; they also lose their capacity to buffer overindulgences at the feeding trough. If that were not enough, this protein affects leptin signalling to promote weight gain and POMC signalling to alter where the fat is stored. Mice without functional sirtuin have larger visceral fat stores.[34] This type of fat produces hormones. As it swells more hormones are produced, altering body chemistry to promote disease. Therefore changes in sirtuin function epigenetically abolish compensatory appetite regulation and promote changes in fat chemistry that contribute to the debilitating diseases associated with obesity.

Sirtuin regulates body weight in response to many signals, one of which is the rate of oxidative damage. When oxidation, due to either pollutants, poisons or energy transfer, is balanced with the availability of anti-oxidants, for example vitamins C and E or phytochemicals, the body is protected against sirtuin-induced unhealthy weight gain. Other anti-oxidants have been shown to stimulate sirtuin's function. These chemicals include quercetin[35] and piceatannol,[36] both of which are typical red wine phenolics, and resveratrol, also found in grapes. However, when the oxidant–anti-oxidant balance is perturbed, the effect is to reduce sirtuin's power to prevent unhealthy body weight. Given that both the rate of oxidation and the requirement for anti-oxidant vitamins increases with fat deposition – the vitamin C is demanded for collagen synthesis and vitamin E to compensate for reserves being sequestered by growing adiposy – it follows that the sector of the population most at risk from body weight gain via this mechanism is the already obese. This risk factor also explains why in the last few decades there have been two new zones tagged onto the BMI. The problem is worsened by a lack of dietary anti-oxidants, nutrients that are usually reserved for high quality foods. People who opt for a nutritionally poor diet are most likely to lack the nutrients needed to protect themselves from the chemical damage caused by housing extra body fat and, via sirtuin, body weight gain is not only linked to calories, but also to anti-oxidants.

Within the cell, the nucleus gains most recognition for housing DNA, and the DNA it houses gains most recognition for determining what we will be. However, there is another sub-cellular compartment that houses DNA: the mitochondria contain a circular nucleic acid thread often overlooked. Yet this smaller piece of DNA has been implicated in the development of many diseases – obesity, diabetes mellitus, neurodegenerative disorders and malignant tumours – and it has been identified as a potential target for many therapies. Because the mitochondria deal in the oxidation process that releases energy from food – they play with fire, so to speak – the DNA in this organelle has a high risk of becoming damaged. In addition to the protective mechanisms in place, there are others that strive to repair damage before it turns into disease. One of these mechanisms involves DNA repair enzymes that scan along the sequence for lesions in the chain and, should they find any, they snip out the offending part and replace it with good stuff. Interestingly, when these repair mechanisms are inactivated in the mitochondria, DNA damage is increased and the body responds by perking up appetite. The mechanism involves changes in the way the body uses genes: mitochondrial DNA damage changes the frequency of the use of chromosomal genes housed in the nucleus. In particular, it affects genes involved in appetite stimulation and those involved in fat synthesis. In contrast, appetite-suppressing genes are used less. Coupled to the increased appetite, the ability to burn fats and make heat is impaired.[62] Whether these changes in gene use represent an adaptive mechanism to combat oxidative damage or whether the corrupted mitochondrial DNA message just causes disease is difficult to say, but such observations do suggest that oxidative damage due to malnutrition or obesity itself could alter appetite control, promoting weight gain. Many chemicals marketed as obesity protectors, for example co-enzyme Q10 and alpha-lipolic acid, are anti-oxidants.

7.4 OBESITY AND AGEING

Increasing sirtuin activity deacetylates the histones, promoting the firm fastening of DNA. This tight wrapping protects against unhealthy body weight gain; it is also suspected to protect against gout, atherosclerosis, diabetes, cancer and Alzheimer's disease.[37] In fact, the sirtuin gene family does not get its fame from its role in body weight regulation, but from its role in regulating ageing, another product of oxidative damage. Indeed, the rate of oxidative damage acts as a timer for age-linked health decline, including the middle-age spread.

In the case of sirtuin's function, ageing and obesity are regulated by energy status and oxidation via histone acetylation. The number of acetyl groups attached to the histones is a function of how rapidly the units are added on and how rapidly they are removed. In both cases, the rate is linked to energy status. The amount of acetylation depends on the amount of available acetyl-co-enzyme A, and therefore on the flow of food energy towards the citric acid cycle. The more energy flowing in that direction, the greater the potential for histone acetylation and the greater the probability that gene use alters to promote ageing and body weight gain. The amount of deacetylation depends on the availability of energy-laden NAD, dubbed NADH, which is in part determined by the energy flow around the citric acid cycle (see The Image of Fat: Molecular Details).

The NAD is charged with energy during the citric acid cycle to become NADH and in this form it acts as sirtuin's co-factor in the removal of acetyl groups from histones. As a result, gene use becomes a function of how rapidly the units are added on and how rapidly they are removed. If the flow of energy towards the citric acid cycle exceeds the rate that it can turn, the amount of acetyl-co-enzyme A may exceed that of NADH, adding more acetyl groups to the histones and, thereby, altering gene use to favour body weight gain. The materials for adiposy are available: the extra molecules of acetyl-co-enzyme A that triggered the change in gene use are the building blocks for fat.

When NADH is charged with energy during the citric acid cycle it can help sirtuin remove the acetyl groups from histones, but when it does, it is cleaved, ending its usefulness. The consumption of NAD links the flux of acetyl groups on and off histones to

micronutrition because vitamin B_3 is one of its precursors. Therefore, energy intake determines the need for this micronutrient:
more food means more acetyl-co-enzyme A and a higher histone
acetylation potential. To maintain a balance, the rate of histone
deacetylation must be raised, consuming more NAD and, indirectly, vitamin B_3. Given that the obese experience a raised metabolic rate they have a higher vitamin B_3 requirement. If this need is
not met the imbalance blocks histone deacetylation and promotes
body weight gain.

It is important to bear in mind that the flexibility in determining
the flux of histone acetylation is not just about promoting body
weight gain or ageing, but is a critical step in regulating the
chemistry of the cell. By measuring the relative amounts of NADH
and acetyl-co-enzyme A the cell can gauge the speed at which the
citric acid cycle turns and it can channel excess energy into fat
stores to prevent a toxic build-up; the relative amounts of acetyl-
co-enzyme A and NADH act as a molecular switch. Once activated, the switch triggers both an adaptation and a chemical
memory: both an instant response and a semi-permanent marking
that could have long-term implications on the fate of acetyl-co-
enzyme A, determining whether it becomes fat or power. Therefore, the factors that affect this signalling – as examples, energy
floods in the form of a supersized cola containing the sugar-
equivalent of a meal; a fatty food feast with the daily calorie
allowance in a single sitting; or a nutritionally poor diet that sees
NADH in short supply (which could involve most of the same
foods as in the first two examples) – alter gene use and leave a
lasting mark on metabolism. These markings may be inherited by
the next generation; parents who flood their bodies with energy to
such an extent that they become obese may not only alter their own
gene use to perpetuate the weight gain cycle but also predispose
their children to the same fate.

Oxidants can alter the balance between NADH and acetyl-co-
enzyme A, promoting epigenetic markings that predestine future
fat. An experience that raises the rate of oxidation, for example
chemical exposure, anti-oxidant deficiencies or energy transfer itself,
could leave the body primed towards fat deposition. The cell can
either use NADH's protective potential against chemical oxidants
lurking about; it can pass on this energy to an enzyme to catalyse a
chemical reaction; it can use the charged molecule to deacetylate

histones; or it can be used to power ATP generation. The cell has a choice in the fate of NADH: power or protection, but not both. Via its various functions, the NAD/NADH ratio acts as a sensor, relaying information about the environment of the cell to alter gene use. The mechanism is automated and devised for survival in the natural world. But sometimes enhancing survival is not the result.

Currently, chemical memories – and also forgotten information – are highlighted as being involved in the development of disease, and so altering epigenetic markings is a logical therapeutic strategy.[38] For this, sirtuin is a current hot target. There are no fewer than seven sirtuin proteins in the cells, tweaking histone deacetylation in response to different signals and in different places along the genome. Blocking sirtuin 1 is thought to be useful for mood stabilisers and anti-epileptics, and a therapy for diabetes,[39] neurodegenerative diseases,[40] HIV[41] and cancer.[42] Some of these treatments are in the clinic now. For example, valproic acid is marketed under various brand names for the treatment of some of these conditions. The mechanism of these drugs is epigenetic.[43] It as though they provide therapy by stimulating gene use and in-house healing pathways that the cells forgot were there – they alter the cells' chemical memories. Yet their use is controversial because this same process leads to ageing and degeneration, triggering some of the diseases the drugs are designed to treat.[44] The obesity-linked member of the family is sirtuin 2. One could imagine that stimulating this enzyme a little may lead to a healthier body weight. Drugs could be designed to compensate for our own stupidity.

Molecular amnesia may be a consequence of diet and a lifestyle that involves high energy assaults, low protection against oxidative damage or micronutrient deficiencies, for example, vitamin B deficiencies affecting NAD synthesis or vitamin C deficiencies limiting protection against oxidation. Via these links, Jirtle may not be far off track in his speculations that obesity protection is linked to good nutrition:[45] many observations point towards this conclusion. Such a molecular mechanism for obesity explains the modern disease: the incidence of the disease; its geographical and socioeconomic distribution; and the increased population of the morbidly obese BMI zone. One could postulate that the rise in incidence of obesity is occurring because the population is generally more exposed to oxidants – for example, as a result of increased calorie consumption, pollution, or a bit of both – and those that resist weight gain do so simply because their nutritional balance offers enough protection to keep the energy balance biased towards a healthy body weight.

7.5 DIETARY YOUTHFULNESS

Diet and lifestyle have been proposed to protect against disease and ageing. When sirtuin and the role of histone acetylation was revealed, a new way of promoting longevity was proposed: calorie restriction.[46] Explained at the most basic level, by forcing our cells to live off the minimum energy possible, the amount of oxidative damage caused as a by-product of energy metabolism is minimised. In addition, by restricting the amount of energy available, histone acetylation is minimised and gene expression is optimised for a healthy body weight and a long life. But the effects are species dependent. Certainly calorie restriction helps mice live longer, and blocking sirtuin 2 activity to increase histone deacetylation has the same effect in yeast.[47] Yet increasing the activity of this same protein in worms does not decrease their lifespan; it has the opposite effect.[48] The interspecies difference comes down to the fact that different species have a different number of sirtuin proteins and we humans have quite a few, making us one of the most complex organisms of all. The difference is also due to the fact that longevity is not just a function of the rate of oxidation, which can be reduced with calorie restriction, but also how rapidly and efficiently oxidative damage is cleared; it is a function of cleaning.

In all species, calorie restriction is thought to protect against ageing by increasing cellular cleaning. Calories are not the only nutrient that can trigger such activity: protein restriction has the same effect. In both cases, the deficit causes the cells to rationalise themselves and sift through their contents, recycling broken and outdated parts so that the resources available best match the cell's current needs. The cleaning involves recycling proteins and organelles and, in the process, decommissions old, worn out or broken parts, including those damaged by oxidation. Whereas protein recycling involves molecular scissors, such as the MMPs, to chop the chains into their amino acid building blocks, organelle recycling is a greater undertaking because of the size and complexity of these sub-cellular compartments. Organelle recycling takes place by a process known as autophagy; the parts are eaten by the cell in a controlled process. The cell breaks down and recycles its own components by enlisting the work of the lysosome organelle. These compartments were discovered by Belgian cytologist Christian de Duve in the 1950s and are essentially cellular stomach sacks containing powerful acid hydrolase enzymes that digest any waste materials and cellular debris that form. They are not fussy about what they eat: food particles, bacteria, viruses and worn-out organelles are all suitable for their digestive processes.

The lysosomes bud off from the Golgi apparatus and embark upon their cleaning missions. Clearly, because of the potential risk of autodigestion, the hazardous work is tightly regulated and SOD enzymes are ready to put out the oxidation fire should it spread outside the area designated by the lysosome bubble membrane. But this process is necessary and plays an important part in cell growth, development and homeostasis, including adaptive processes that redistribute resources in times of need. A hungry cell may digest surplus proteins and other parts of itself to reallocate resources from non-essential processes to those that sustain life.

In order to digest parts of the cell, they must be on the inside of the lysosome bubble. The mechanism of internalisation depends on the size of the meal. A little bit of protein, for example pieces that have not folded correctly during their synthesis, can directly translocate across the lysosome membrane. Whole proteins are too big to enter via this route, so are internalised by creating a pit on the membrane surface just big enough to accommodate the protein; a process called invagination. The membrane pocket is then closed so the protein is held in a membrane-bound bubble on the inside of the lysosome. At the right moment, the bubble bursts, tipping the protein into the acid bath. Organelles need a different mechanism of autophagy, principally because they are often larger than the lysosome itself. In this case, a double membrane wraps around the whole organelle scheduled for decommissioning to make a temporary structure known as an autophagosome. Its sole function is to deliver the cargo to the lysosome. The two bodies fuse and the contents of the lysosome set to work eating the organelle. Once proteins have been digested into amino acids, fats into fatty acids and complex carbohydrates into sugars, these building blocks can be reused by the cell. Any materials that are too badly damaged for recycling can be flagged by a molecular marker for export and excretion.

Certain metabolic changes, including those involved in disease progression and ageing, are associated with decreased autophagy. Unhealthy fat deposits can slow autophagy down, an effect mediated via the sirtuin proteins.[49] Age and adiposity are both characterised at a cellular level by the increased occurrence of incomplete autophagy of the mitochondria; an accumulation of over-sized and inefficient versions of these organelles in the cells. Although the total metabolic rate in the obese seems to be raised rather than reduced, the efficiency at a molecular level is compromised because autophagy has been lazy and the cell is not cleaned. And that is not all. Without efficient cleaning, inactivated proteins accumulate; they are inactivated as a result of oxidation and poisoning. Carbon monoxide is one such poison. Carbon monoxide is a prime example of a cell signaller with dose-dependent effects. It is naturally produced in small amounts by the body, but too much binds tightly to proteins, blocking their function. It is most famed for its asphyxiating interaction with haemoglobin, but it also blocks other proteins, including cytochrome c oxidase, an important protein in the energy release process. Carbon monoxide makes cells sluggish. In many studies, protein or energy restriction – not necessarily starvation, but a fast[50] – seems to increase organelle recycling and boost health, removing both the over-sized mitochondria and the inefficient carbon-monoxide-bound proteins they contain.

Fasting is part of many ancient traditions, but the healthy application does not necessarily mean avoiding all food; it can be a change of dietary habits that reduces the amount of fat and protein for a period of time. Consider the effect of short-term protein restriction: it seems to alter body chemistry to increase the amount of thyroid hormone signalling; reduce the amount of neuropeptide signal dulling induced by somatostatin; and it may also reduce the advance of the ageing process by promoting spring cleaning.

Mice fed on protein-restricted diets are leaner[51] and live longer[52,53] than mice fed on a regular diet, and the reason for these apparent health benefits comes down to changes in body chemistry involving the recycling of amino acids. When the cell has fewer amino acids than it needs, it begins a remodelling programme dependent on autophagy. These processes not only act as a molecular buffer against shortages, but also help to prevent degeneration and ageing. In balancing the body's needs to cope with a protein crisis, defunct and less efficient organelles are inadvertently decommissioned, improving the vitality of the cell. So important is this process that reduced autophagy efficiency causes degeneration; frailty – skeletal muscle weakening – does not just involve reduced muscle mass, but also changes in the chemistry of the cells. Super-sized but inefficient mitochondria appear in cells with age, and once formed these large organelles are difficult to clear. As time progresses, more and more damaged or non-functional macro-molecules and organelles accumulate.[54] This molecular garbage is linked to general degeneration and many diseases.[55] And so, preserving youthfulness is more than skin deep; it extends down to molecular details.

Exercise can maintain molecular youth. In addition to increasing muscle strength in older adults, exercise also appears to reduce the evidence of oxidative stress and improves mitochondrial func-tion.[56] The function is measured collectively, implying that the inefficient members of this group of organelles are decommis-sioned, probably following signals for muscle building. The increased protein demand raises the rate of protein turnover such that cells actively spring clean in search of worn out parts – molecular garbage that can be recycled.

Body weight control is another method of preserving youthful-ness: among the obese, the rate of oxidative damage is accelerated compared with people of a healthy weight.[57] Accelerated oxidation could have a number of causes. First, the epigenetically raised MMP9 levels snipping away at collagen soak up the stocks of vitamin C, leaving less to protect the cells. Secondly, overall the obese burn more fuel, generating more reactive oxygen species, and therefore require more micronutrition to obtain the same level of protection as the slim. In both cases, ageing and degenerative disease are promoted and at a molecular level obesity can be defined as a state of high oxidative stress,[58] which is a harbinger of diseases: diabetes is one,[59] cancer is another;[60] and unhealthy weight gain is a third, generating a vicious cycle. The accelerated rate of oxidation extends outside the cells and into the blood as body fat alters the lipid profile.[61] Circulating high density lipo-protein (HDL) mops up excess cholesterol from the body and ships it back to the liver for processing. These assemblies also have an anti-oxidant and anti-inflammatory role that is modulated by the paraoxonase-1 (PON1) enzyme on their surface, but the efficiency of this protection depends on the BMI. The ability of HDL to protect against oxidative damage is inversely correlated with fat deposits: as the BMI rises, PON1 is less able to protect against damage, which is thought to accelerate the development of diseases associated with high body weight, including cancer and arteriosclerosis.[61]

In the case of obesity, the oxidative damage is a vicious cycle. This chemical activity triggers body weight gain, which in turn triggers further oxidative damage. The higher levels of oxidation and the altered epigenome associated with obesity demonstrate that this disease is not simply a question of padding, but a fundamental change in body chemistry that triggers a series of serious disease states. Concern about rising body weight is not rooted in vanity, but is a prudent way of protecting the body against degenerative disease. It is a health investment that improves the length and quality of life. The self-promoting nature of the disease warrants vigilance for the early signs of weight gain. Part of the vigilance should focus on the wider aspects of nutrition: not just the calories, but the way they are delivered; not just the calories, but the micronutrient balance. So many nutrients are important for healthy body weight regulation, all at the correct dose.

REFERENCES

1. A. Bishayee and M. Chatterjee, *Biol. Trace Elem. Res.*, 1995, **48**, 275–285.
2. A. M. Hodge, D. R. English, K. O'Dea and G. G. Giles, *Am. J. Epidemiol.*, 2007, **165**, 603–610.
3. J. L. Nadler, T. Buchanan, R. Natarajan, I. Antonipillai, R. Bergman and R. Rude, *Hypertension*, 1993, **21**, 1024–1029.
4. T. W. Balon, A. Jasman, S. Scott, W. P. Meehan, R. K. Rude and J. L. Nadler, *Hypertension*, 1994, **23**, 1036–1039.
5. R. Swaminathan, *Clin. Biochem. Rev.*, 2003, **24**, 47–66.
6. G. C. Major, J.-P. Chaput, M. Ledoux, S. St-Pierre, G. H. Anderson, M. B. Zemel and A. Tremblay, *Obes. Rev.*, 2008, **9**, 428–445.
7. F. G. Hopkins, *J. Physiol. (Lond.)*, 1912, **44**, 425–460.
8. S. H. Wei, V. Episkopou, R. Piantedosi, S. Maeda, K. Shimada, M. E. Gottesman and W. S. Blaner, *J. Biol. Chem.*, 1995, **270**, 866–870.

9. D. Yeste, J. Vendrell, R. Tomasini, L. L. Gallart, M. Clemente, I. Simon, M. Albisu, M. Gussinye, L. Audi and A. Carrascosa, *Horm. Res. Paediatr.*, 2010, **73**, 335–340.
10. C. F. Garland, F. C. Garland and E. D. Gorham, *Am. J. Clin. Nutr.*, 1991, **54**, S193–S201.
11. S. Masuda and G. Jones, *Mol. Cancer Ther.*, 2006, **5**, 797–808.
12. H. Yao, W. Xu, X. Shi and Z. Zhang, *J. Environ. Sci. Health, Part C – Environ. Carcinog. Ecotoxicol. Rev.*, 2011, **29**, 1–31.
13. K. S. Solanky, N. J. Bailey, B. M. Beckwith-Hall, S. Bingham, A. Davis, E. Holmes, J. K. Nicholson and A. Cassidy, *J. Nutr. Biochem.*, 2005, **16**, 236–244.
14. M. L. Ricketts, D. D. Moore, W. J. Banz, O. Mezei and N. F. Shay, *J. Nutr. Biochem.*, 2005, **16**, 321–330.
15. P. McCue, Y. I. Kwon and K. Shetty, *J. Food Biochem.*, 2005, **29**, 278–294.
16. M. M. Huttunen, P. E. Pietila, H. T. Vijakainen and C. J. E. Lamberg-Allardt, *J. Nutr. Biochem.*, 2006, **17**, 479–484.
17. D. Mason, L. Suva, P. Genever, A. Patton, S. Steuckle, R. Hillam and T. Skerry, *Bone*, 1997, **20**, 199–205.
18. M. G. Packard and S. A. Chen, *Psychobiol.*, 1999, **27**, 377–385.
19. H. Kaba, Y. Hayashi, T. Higuchi and S. Nakanishi, *Science*, 1994, **265**, 262–264.
20. C. H. Park, S. H. Choi, Y. Piao, S.-H. Kim, Y.-J. Lee, H.-S. Kim, S.-J. Jeong, J.-C. Rah, J.-H. Seo, J.-H. Lee, K.-A. Chang, Y.-J. Jung and Y.-H. Suh, *Toxicol. Lett.*, 2000, **115**, 117–125.
21. L. M. Robison, D. A. Sclar, T. L. Skaer and R. S. Galin, *Clin. Pediatr.*, 1999, **38**, 209–217.
22. B. Kaplan, J. McNicol, R. Conte and H. K. Moghadam, *Pediatrics,* 1989, **83**, 7–17.
23. T. N. Huang and S.-W. Ching, *The Yellow Emperor's Classic of Internal Medicine*, University of California Press, Los Angeles, California, 1996.
24. J. Wortsman, L. Y. Matsuoka, T. C. Chen, Z. R. Lu and M. F. Holick, *Am. J. Clin. Nutr.*, 2000, **72**, 690–693.
25. L. Newbury, K. Dolan, M. Hatzifotis, N. Low and G. Fielding, *Obes. Surg.*, 2003, **13**, 893–895.
26. M. V. Maffini, B. S. Rubin, C. Sonnenschein and A. M. Soto, *Mol. Cell. Endocrinol.*, 2006, **254**, 179–186.
27. D. C. Dolinoy, D. Huang and R. L. Jirtle, *Proc. Nat. Acad. Sci. USA*, 2007, **104**, 13056–13061.

28. J. Cox, S. Williams, K. Grove, R. H. Lane and K. M. Aagaard-Tillery, *Am. J. Obst. Gynecol.*, 2009, 201.
29. F. L. Muller, M. S. Lustgarten, Y. Jang, A. Richardson and H. Van Remmen, *Free Radical Biol. Med.*, 2007, **43**, 477–503.
30. T. Kouzarides, *Curr. Opin. Genet. Dev.*, 2002, **12**, 371–371.
31. C. A. Perry, C. D. Allis and A. T. Annunziato, *Biochem.*, 1993, **32**, 13615–13623.
32. P. Jeppesen, *Bioessays*, 1997, **19**, 67–74.
33. J. Campion, F. I. Milagro and J. A. Martinez, *Obesity Rev.*, 2009, **10**, 383–392.
34. G. Ramadori, T. Fujikawa, M. Fukuda, J. Anderson, D. A. Morgan, R. Mostoslavsky, R. C. Stuart, M. Perello, C. R. Vianna, E. A. Nillni, K. Rahmouni and R. Coppari, *Cell Metab.*, 2010, **12**, 78–87.
35. L. Palmieri, M. Mameli and G. Ronca, *Drug. Exp. Clin. Res.*, 1999, **25**, 79–85.
36. K. T. Howitz, K. J. Bitterman, H. Y. Cohen, D. W. Lamming, S. Lavu, J. G. Wood, R. E. Zipkin, P. Chung, A. Kisielewski, L. L. Zhang, B. Scherer and D. A. Sinclair, *Nature*, 2003, **425**, 191–196.
37. J. D. Adams, Jr. and L. K. Klaidman, *Lett. Drug Des. Discov.*, 2007, **4**, 44–48.
38. N. Batty, G. G. Malouf and J. P. J. Issa, *Cancer Lett.*, 2007, **280**, 192–200.
39. J. C. Milne, P. D. Lambert, S. Schenk, D. P. Carney, J. J. Smith, D. J. Gagne, L. Jin, O. Boss, R. B. Perni, C. B. Vu, J. E. Bemis, R. Xie, J. S. Disch, P. Y. Ng, J. J. Nunes, A. V. Lynch, H. Yang, H. Galonek, K. Israelian, W. Choy, A. Iffland, S. Lavu, O. Medvedik, D. A. Sinclair, J. M. Olefsky, M. R. Jirousek, P. J. Elliott and C. H. Westphal, *Nature*, 2007, **450**, 712–716.
40. E. Hahnen, J. Hauke, C. Tränkle, I. Eyüpoglu, B. Wirth and I. Blümcke, *Expert Opin Investig. Drugs*, 2008, **17**, 169–184.
41. http://www.clinicaltrials.gov/ct2/show/NCT00576290?term=HDAC&rank=8 (accessed on 19 August 2011).
42. http://www.drugs.com/nda/zolinza_060607.html (accessed on 15 May 2012).
43. C. Monneret, *Anti-Cancer Drugs*, 2007, **18**, 619–619.
44. I. van Leeuwen and S. Lain, *Adv. Cancer Res.*, 2009, **102**, 171–195.
45. R. A. Waterland and R. L. Jirtle, *Mol. Cell. Biol.*, 2003, **23**, 5293–5300.

46. W.-L. Yen and D. J. Klionsky, *Physiol.*, 2008, **23**, 248–262.
47. M. Kaeberlein, M. McVey and L. Guarente, *Genes Dev.*, 1999, **13**, 2570–2580.
48. H. A. Tissenbaum and L. Guarente, *Nature*, 2001, **410**, 227–230.
49. I. H. Lee, L. Cao, R. Mostoslavsky, D. B. Lombard, J. Liu, N. E. Bruns, M. Tsokos, F. W. Alt and T. Finkel, *Proc. Nat. Acad. Sci. USA*, 2008, **105**, 3374–3379.
50. S. Nemoto, M. M. Fergusson and T. Finkel, *Science*, 2004, **306**, 2105–2108.
51. P. C. Lisbo, M. C. F. Passos, S. C. P. Dutra, R. S. Santos, I. T. Bonomo, A. P. Cabanelas, C. C. Pazos-Moura and E. G. Moura, *J. Endocrinol.*, 2003, **177**, 261–267.
52. R. A. Miller, G. Buehner, Y. Chang, J. M. Harper, R. Sigler and M. Smith-Wheelock, *Aging Cell*, 2005, **4**, 119–125.
53. M. D. W. Piper, W. Mair and L. Partridge, *J. Gerontol. A*, 2005, **60**, 549–555.
54. A. Terman and U. T. Brunk, *Exp. Gerontol.*, 2004, **39**, 701–705.
55. T. Hara, K. Nakamura, M. Matsui, A. Yamamoto, Y. Nakahara, R. Suzuki-Migishima, M. Yokoyama, K. Mishima, I. Saito, H. Okano and N. Mizushima, *Nature*, 2006, **441**, 885–889.
56. M. A. Tarnopolsky, *Appl. Physiol. Nutr. Metabol.*, 2009, **34**, 348–354.
57. M. E. Atabek, M. Keskin, C. Yazici, M. Kendirci, N. Hatipoglu, E. Koklu and S. Kurtoglu, *Eur. J. Pediatr.*, 2006, **165**, 753–756.
58. J. V. Higdon and B. Frei, *Arterioscl. Thromb. Vasc. Biol.*, 2003, **23**, 365–367.
59. G. E. Striker and H. Vlassara, *US Endocrinol.*, 2010, **6**, 14–19.
60. H. A. Al-Aubaidy and H. F. Jelinek, *Eur. J. Endocrinol.*, 2011, **164**, 899–904.
61. G. Ferretti, T. Bacchetti, S. Masciangelo and V. Bicchiega, *Obesity*, 2010, **18**, 1079–1084.
62. X. C. Zhang, H. J. Spencer, C. Zuo, M. Higuchi, G. Ranganathan, P. A. Kern, M. W. Chou, Q. Huang, B. Szczesny, S. Mitra, A. J. Watson, G. P. Margison and C. Y. Fan, *Am. J. Pathol.*, 2011, **178**, 1715–1727.

CHAPTER 8

A Fat Future?

When Harvey Cushing was cataloguing the symptoms of a new syndrome between the 1930s and the 1950s, he studied the then most common cause – the hardwired, genetic disease. His work was on a recessive trait triggering tumour growth on the pituitary glands that causes the over-production of adrenocortico-trophic hormone, ACTH (see Stone Age Obesity: Of Mice and Men) and, in turn, pushes up cortisol levels. Cortisol, a gluco-corticoid synthesised from cholesterol, is an important hormone in growth and development. The amount of this signaller circulating in the body fluctuates throughout life according to diet, lifestyle and health. For example, women in the last trimester of pregnancy generally have higher levels of cortisol than at other times in their lives. Its function in pregnancy is to promote weight gain in mother and child. It achieves this effect by counterbalancing insulin and, therefore, may be behind the pre-diabetes and insulin resistance common during gestation (see Beyond the Helix: Sacrifices for our Children). There are other triggers for cortisol release including diet, sleep deprivation, disease and stress. In each case, cortisol alters body chemistry in an attempt to deal with the challenge and when the urgency is over, the system re-equilibrates at a healthy level without lasting effects. If the cause is hardwired, as in the case of Harvey Cushing's patients, signalling levels are higher for longer and disease results.

Fat Chemistry: The Science behind Obesity
Claire S. Allardyce
© Claire S. Allardyce 2012
Published by the Royal Society of Chemistry, www.rsc.org

Cushing's syndrome is an endocrine disease: a disease that involves glands and hormone signallers. Such diseases are powerful and have far reaching consequences. In this case, some of the symptoms include reduced bone formation, insulin resistance and fat deposition on the midriff; a location where fat produces hormones like a gland. As the abdominal fat gland swells, the body implements mechanisms to prevent a further health decline by actively excreting cortisol in the urine.[1] This strategy may work as a short-term fix, but with continued spikes the excretion creates imbalances among other hormones and leads to other health consequences. Similarly, the raised ACTH production in the type of Cushing's syndrome the great man himself studied triggers additional symptoms, including altered energy metabolism and increased weight gain. In the case of Cushing's syndrome, weight gain is a sign of systemic disruption to fat chemistry caused by disruptions to hormone signalling; it is a very observable symptom of a far reaching disease state.

Since Cushing's founding work, his name has been used as an umbrella term to categorise elevated cortisol levels, whatever the cause. Increasingly the cause is environmental. The outward signs are identical, including obesity due to an overwhelming appetite and fat distribution to promote a thickening waistline, but despite the same need to feed, the two causes differ in their curability. The hardwired version requires medical intervention – surgery, radiation therapy, medication or a combination of the three – but the environmentally induced equivalent can be readily cured by changes in diet and lifestyle. Over the past decades, the incidence of the hardwired version of the disease has remained the same, but that of the environmentally induced version has begun to rise. A few decades into the future there may be an epidemic of the full blown disease, but at the moment it is still rare. It is not a major contributor to the modern obesity problem yet the rise in the incidence of Cushing's may be a reflection of a general rise in cortisol levels, too mild to cause the syndrome, but perhaps significant enough to contribute to the weight gain we love to hate.

Milder increases in signalling can cause milder symptoms. Such disease states are not classifiable as Cushing's, but they are powerful enough to trigger changes in body weight. Known culprits for raising cortisol include medicines such as oral contraceptive pills and steroid treatments for rheumatoid arthritis, inflammatory bowel disease or asthma, and both types of drug have become more widely utilised with the rise of the obesity epidemic. The prognosis for the future is that, unless there is dietary change, the use of these drugs will continue to rise: dads with a taste for fatty foods produce children with an altered epigenome, including both metabolic disturbances and increased use of the *Il13ra2* gene, which is linked to allergies, asthma and allergic lung disease (see Beyond the Helix: Sacrifices for our Children).

Lifestyle has also changed to promote cortisol production. High stress lifestyles can trigger the release of more of this hormone than sedate living. The experience can be annotated onto the epigenome (see Beyond the Helix) and inherited. Stress increases methylation on the *NR3C1* gene, the gene that encodes a cortisol-sensitive receptor,[2] such that cortisol levels must be raised to achieve the normal level of signalling. Cortisol resistance is a disease state and can be localised, causing asthma for example, or generalised. In the latter case, knock-out mice lacking the ability to respond to cortisol die of respiratory distress syndrome within a few hours of birth.

The types of stress that increase cortisol production can be physical, as experienced by highly trained athletes, or mental. In particular, people suffering from depression, alcoholism or panic disorders produce more cortisol than their healthier peers. This change in body chemistry is accompanied by many other changes that perturb mental health, mood and body weight. It is widely suggested that modern, high stress lifestyles could contribute towards elevated levels of this signaller. This profile promotes fat storage on the midriff and, once stored there, the fat begins to function as an endocrine gland, releasing hormones and signallers that change the balance of fat chemistry. Therefore, stress, via cortisol, could promote the biochemical changes that cause the health decline associated with obesity even before the sign of unhealthy body weight is apparent.

In 1972, two ophthalmologists, Ralph Levene of New York University School of Medicine and Bernard Schwartz of Tufts New England Medical Centre, Boston, teamed up with an anthropologist, Peter Workman of the University of Massachusetts, to investigate claims that cortisol levels were genetically determined. They found no evidence of heritability, but they did find an unexpected yet significant correlation in the cortisol levels of husband and wife pairs.[3] These data suggest that cortisol levels are strongly influenced by the shared environment of the spouses. By 1988, the heritability of plasma cortisol levels was determined to be 50%,[4] and in 2003 the figure was revised and increased.[5] The authors of this later report, psychologists interested in the pivotal role of cortisol in mental and physical health, suggested that the previous estimates were flawed because the researchers did not appropriately delineate the contributions of genetic and environmental factors. Their quoted figure for the heritability of cortisol levels was 62%, an estimate based purely on twin studies in an attempt to remove some of the genetic variability. They inadvertently removed some of the most critical environmental variability too, because these individuals shared the same perinatal environment and are likely to have shared the same perinatal imprinting. Therefore, in addition to the genome determining heritability, it seems as though the epigenome could have an effect too; an environmentally induced effect. One could also hypothesise that the observed increase in heritability of cortisol levels since 1972 may be due not only to revised methods of data analysis, but to environmental change, increasingly priming the population for elevated levels of this signaller. The *in utero* stress experienced by the individuals with increased *NR3C1* glucocorticoid receptor gene methylation is linked to maternal depression and anxiety – mental health issues elevated the mums' cortisol levels and in so doing affected the health of their babies. Therefore stress can be added to the list of parental experiences that mould our metabolism.

8.1   THE INTERTWINED NATURE OF FOOD AND MOOD

Stress can take on many forms. For the Stone Agers, stress was mostly physical: snake bites, parasitic worms, frostbite and hunger, whereas modern humans experience hyper-stress, most easily defined as the tension caused by the sheer busyness of life, for example juggling work and family commitments; the severity of the consequences of decisions in the work environment, from life and death medical decisions through to financial decisions that affect the jobs and security of many people; constant financial pressure; or simply working in a fast paced environment. We are Stone Agers living in the fast lane.

Given that stress is linked to appetite and metabolism, it has become a popular target for blame as the cause of unhealthy weight gain. In addition to the way cortisol affects fat distribution, stress can manipulate levels of other hormones to affect body weight. The hardwired disease Cushing studied involves raised levels of ACTH pushing up cortisol levels; effectively it is a disease that triggers the biochemical stress response without a valid trigger. The consequences – the symptoms of Cushing's disease – are the same as those of too much stress.

The symptoms of stress are diverse, affecting mood and metabolism. These features are fundamentally interlinked by the nature of the gene that encodes ACTH: the *POMC* gene site (see Stone Age Obesity: Obesity Resistance). It is via the POMC neurons that leptin mediates part of its effect, and via the POMC signaller sequence that sirtuin alters fat deposition. Via this lone signaller, energy balance can be biased towards body weight gain; changes in fat deposition can induce chemical obesity at a lower than expected BMI; and compensatory appetite regulation can be lost – all the features of the modern obesity epidemic can be explained. These wide reaching effects are due to the fact that *POMC* is not just the gene for ACTH. The product of its translation is a string of amino acids – or a polypeptide, as it is more commonly known in scientific circles. Rather than forming a protein, this string just sits around and waits. The polypeptide signaller is waiting for the right stimulus to determine its fate, because within this string are no fewer than eight signallers, each with a different function that collectively intertwine diet, lifestyle and stress with appetite and the amount of stored body fat and its distribution. Food has long been linked to mood, but the degree of intricacy in the relationship is only just beginning to be understood.

The eight signallers in the POMC sequence have various roles in adaptive responses to environmental change, and their triggers extend from sunshine through to stress. Because the peptide also includes signallers that can influence fat chemistry, these same triggers can influence body weight. At the first level, the influence is a transient adaptation that biases the energy balance. However, some of the adaptations involve changing the way fat is stored: physiological consequences with a longer duration and knock-on effects that may be recorded in the genes and inherited by the next generation.

The POMC signallers are not tethered one after another in a sequential arrangement, but the beginnings of some overlap with the ends of others such that their excision is interdependent.

Starting simply, the first signaller does not overlap with the others. Its name is gamma-melanocyte-stimulating hormone (g-MSH) and it is composed of just 12 amino acid building blocks. It is too small to be classed as a protein; it is a peptide signaller. Its roles are in regulating blood pressure and sodium excretion by the kidneys, and its excision from the POMC polypeptide is dictated by the amount of sodium in the blood,[6] which in turn is a function of salt intake. Therefore, dietary salt has a role in dictating the fate of POMC and there is evidence to suggest that it may also affect the production of the signaller: salt appears to reduce POMC production,[7] increasing appetite.

After g-MSH, two signallers begin simultaneously in the POMC sequence: ACTH and alpha-melanocyte-stimulating hormone (a-MSH). The excision of these two signallers is mutually exclusive, but in addition, because the two signallers start at the same place in the peptide, the smaller forms part of the larger and is released during its degradation process: when time is up for ACTH it is snipped to release a-MSH. Given that one signaller is part of the other, increasing the amount of ACTH signalling also increases the amount of a-MSH. The smaller signaller is an appetite suppressant and is also involved in the body's damage limitation response to sodium excess, preventing salt-sensitive hypertension[8] whilst inducing changes in opioid signal responses in the brain and relaxing the rectum;[9] the consequences of this can be beneficial or embarrassing. Therefore, high salt diets that reduce POMC production could increase the risk of constipation whilst increasing the body's sensitivity to sodium – another vicious cycle of declining health.

POMC is produced in the anterior lobe of the pituitary gland, where stress signals trigger ACTH to be excised and released into the blood. ACTH survives for about 10 minutes, making its way down to the adrenal glands that sit upon the kidneys. There it binds to its receptor switch on the adrenocortical cells to trigger the synthesis and secretion of steroid signallers. The knock-on effect of this signal is that cholesterol is taken up for steroid synthesis, triggering another wave of downstream signalling. The steroids synthesised in reseponse to the ACTH signal are lipids made up of 17 carbon atoms arranged in four rings with various other clusters of atoms docked on. The family of steroid signallers includes bile acids, sex hormones, vitamin D and the adrenal hormones, of which cortisol is the main steroid produced in response to ACTH; bursts of cortisol occur in response to stress.

When ACTH is degraded, it releases corticotropin-like intermediate peptide (CLIP) and a-MSH; a further wave of signalling is generated. CLIP increases the release of insulin from the pancreas;[10] it signals for energy storage. It also stimulates the release of other hormones, including starch-digesting amylase, suggesting that the energy storage is most likely to occur as fat. Like cortisol, a-MSH has some very different functions: it is responsible for skin tanning, it regulates appetite and it has been implicated in causing sexual arousal; at least, synthetic analogues of this signaller caused enhanced libido and erections in some men. This signaller triggers tanning by relaying messages from one cell to another. Keratinocytes, which form the majority of the part of the skin we see, place a take away order with melanocytes located nearby for the melanin pigment that darkens skin, and the messenger that carries the order is a-MSH. When ultraviolet light strikes, the cells increase POMC production and cleave the precursor to release a-MSH, which is then secreted so that it can travel to nearby melanocytes where it signals for the production of the melanin pigment. In turn, the melanin is secreted by the melanocytes and taken up by the keratinocytes. For other animals, altered production of a-MSH is a matter of life and death. The clawed toad *Xenopus laevis* outwits its predators by camouflaging itself in shady corners using rapid, MSH-induced skin darkening. Apart from tanning, it is not a major cause of colour variation in humans, except darkening during pregnancy and inherited, severely pale colouring caused by polymorphisms that impede its production or function. The pale colouring is sometimes accompanied by body weight gain because this signaller is also an appetite quencher – watch out, Viagra! Here is a potential combined therapy for erectile dysfunction and obesity in one. Indeed, because the production of the natural form of a-MSH is stimulated by sunshine, it may be responsible for reduced appetite during the longer daylight hours and summer love.

After ACTH (or the combination of a-MSH and CLIP) two more signallers are contained in the POMC sequence, again starting together so their excision is mutually exclusive; these are the lipotropins. Beta-lipotropin runs the full length of the remaining POMC sequence. Just like a-MSH, beta-lipotropin has effects on pigmentation and appetite. In the skin, this signaller stimulates melanin production, and elsewhere it triggers lipid-mobilisation, lipolysis and steroidogenesis; lipotropin promotes fat burning. Once again, the effect of light is to promote a balance of fat chemistry that trims and tones the body.

When beta-lipotropin is clipped in the right place it produces gamma-lipotropin and beta-endorphin. Like beta-lipotropin, gamma-lipotropin mobilises fats for metabolism. There is a third signaller that can be produced from the gamma-lipotropin part of the sequence, called b-MSH. The role of b-MSH seems to be much more powerful and more direct: this signaller plays a critical role in body weight regulation such that changes in the POMC gene that abolish the function of this signaller lead to early onset obesity.[11]

The roles of beta-endorphin in body chemistry are much more complicated because this signaller has many functions, but not in the domain of pigmentation. In the brain and nervous system, it is an endogenous opioid peptide neurotransmitter released into blood from the pituitary gland. It is also produced in the hypothalamic neurons where it is released directly into the spinal cord and brain. These two routes do not overlap, because the signaller cannot cross the blood–brain barrier and, whereas its role in the blood is not clear (although there is evidence to suggest it may dull pain),[12] in the nervous system it binds to opioid receptors. It has the highest affinity for the same receptor as morphine, yet the natural version has in the region of an 80-fold higher analgesic potency compared to the external source. Indeed, possibly the most concrete function of this signaller is in pain relief, mediated by inhibiting neuro-transmitter release and increasing dopamine release; its place in POMC suggests that it may reduce the sting of sunburn when it is produced as a by-product in the tanning process.

Stimulators or mimics of beta-endorphin have many medical applications because of their role in regulating cardiac, gastric and vascular function, as well as possibly panic and satiation. It has also been suggested to slow the growth of cancer cells (keep in mind the fact that table salt reduces the amount of the beta-endorphin precursor, POMC, and sunshine perks it up). Indeed, beta-endorphin is supposed to be involved in the chemical mechanism behind the success of acupuncture.[13] Chemical stimulation of the endorphins using drugs, such as naltrexone, has been trialled as a treatment involved in pain management in many diseases including multiple sclerosis,[14] Crohn's disease,[15] HIV/AIDS[16] and neurodegeneration.[17]

In the future, there may also be a treatment for obesity in there too: this signaller decreases the production of gastric juices, taking our minds off food.[18] So natural endorphins seem to promote health and could help weight control, but how do we encourage our bodies to produce more of these signallers? The answer is with a healthy lifestyle. Endorphins are released during both endurance exercise and relaxation;[19] spend this time outside and the positive effects of sunshine may stimulate endorphin production a little more.

When genetic polymorphisms affect the work of the ACTH signaller, their cortisol-boosting effects can induce Cushing's syndrome; this hardwired disease attracted the great man's attention. Today the most common cause is the environment. Because this disease works via multi-factorial POMC signalling, the environmental disease may have one of many causes: medicines, diet, stress, exercise and sunshine are involved in regulating the signalling of this sequence. It follows that perturbations in any one could promote body weight gain; any combination of such perturbations may combine to give a stronger effect. Via POMC signalling, an unhealthy lifestyle can initiate a series of detrimental effects – effects that touch many areas of our physical and mental well-being. Via POMC, environmental triggers affect fat chemistry in a transient and adaptive way. Via POMC signalling, stress seems a likely contender for obesity.

8.2 THE MEMORY OF MALNUTRITION

There is a type of stress that deserves a section to itself, and that is malnutrition. The most readily recognisable type of malnutrition is what we term protein–energy malnutrition or starvation, and it is usually the result of a lack of food in general. Although all nutrients are needed to maintain good health, protein and energy are needed in the greatest amounts, hence the name. Marasmus is the term used to describe severe emaciation to the extent that there is extensive tissue and muscle wasting. The term is derived from the Greek word meaning starvation. Stress caused by such lack of food unsurprisingly induces mood changes, including fretful and irritable behaviour and incredible hunger. Some of these observations are a result of cortisol signalling. Although hunger and signals for fat stockpiling have no effect on body weight unless there are resources available, these experiences can be imprinted onto the genome to give lifelong consequences.

In 2004 and 2005 Luciane Peter Grillo, of the Universidade do Vale do Itajaí, Brazil, published reports that linked childhood nutritional shortages to slower metabolism and, thus, weight gain in adulthood. Grillo showed that the bodies of children stunted by malnutrition are thrifty with their supplies and less willing to tap into fat reserves than those of children who have not suffered from shortages.[20] They did not inherit genes that predisposed them to a life of obesity; rather, they inherited the same variety of genetic polymorphisms abundant in the general population, but environmental stress – malnutrition – changed the volume control on some genes to promote weight gain via multiple mechanisms, including conserving energy. When daily energy use is corrected for body size, the energy requirements of those stunted as a result of malnutrition can be directly compared with those of the average person, and it has been found that during rest the nutritionally stunted use 4% less energy. This amount may not seem significant, but over a lifetime, it can reduce energy requirements sufficiently to cause unhealthy body weight gain. This link was enforced by researchers from the Clinical Diabetes and Nutrition Section, National Institute of Diabetes and Digestive and Kidney Diseases, Phoenix, Arizona, who used a respiratory chamber to measure the energy expenditure of a group of Pima Indians and then monitored their rate of change in body weight over a two-year period. Those with the lowest energy expenditure had a four-fold higher risk of gaining weight compared with the people who expended more.[21] That body weight increased implies that the people ate more than they needed; compensatory appetite regulation was also compromised.

Resting energy use is not the only difference in fat chemistry that operates after experiencing times of nutritional hardship: when fasting, the nutritionally stunted burn fats to produce energy at about two-thirds the normal rate. Thus, after experiencing sufficient food shortages to affect growth, gene use allows survival on much less energy than someone of the same height and build who has never experienced these shortages: an environmentally induced thrifty epigenotype.

Stunting has been highlighted as a strong contributor to obesity in developing countries. Take Chile as an example. As many forms of malnutrition decreased during the 1980s, due to a generally improved and more varied diet, obesity began to rise rapidly in all age groups and the portion sizes of food aid have been strongly implicated in promoting the unhealthy weight gain. The high-energy snack foods that were provided for these people in an attempt to reverse the trend of undernutrition and starvation were based on the recommended portion for a well nourished and significantly taller person, resulting in overeating among the stunted.[22] In the same period, fast food was introduced in the internationally standard portion sizes.

In addition to nutritional shortages that cause anatomical changes, such as marasmus and stunting, less severe shortages may still trigger imprinting that promotes obesity. Juan Manuel Rios-Torres, of the Endocrinology and Metabolism Department at the INCMNSZ in Mexico City, found that protein–energy malnutrition in early life – not starvation, just shortages – is linked to insulin insensitivity and midriff weight gain in adulthood. Instead of these problems leading to obesity, it seems they follow excessive weight gain.[23] His subjects' bodies had memories of their nutritional experiences, but did not remember the lessons learned until body weight rose. And when they finally recollected, the effect was devastating.

Those who are stunted are not the only people to undergo a two stage mechanism that leads to obesity: the pattern seems to be universal and adds a further level of complexity in understanding the origin of the epidemic. However, the effect of protein–energy malnutrition in reducing the overall metabolic rate does not seem to apply as a cause of the global obesity epidemic. The current availability of dietary protein and energy is at an all-time high, and so is the height of the population. There have also been a vast number of studies gauging metabolic rates amongst individuals, and the data seem to suggest that, certainly in industrialised nations, the obese do not burn less energy than the trim. On average, the overall amount of energy burnt by an individual is about the same, irrespective of the human development index of the country.[24] Many of these studies were performed at a molecular level. Dale Schoeller applied his nifty double-labelled water studies beyond probing weight loss on low-carbohydrate diets (see Left to our own devices: Calories don't count, at least not proportionally), and what he found was that, overall, the obese are burning more fuel than their trim peers.

Total metabolism can be subdivided into two categories: resting energy expenditure (REE) and activity energy expenditure (AEE). Both values are rates: measures of energy transformation with time. The REE is a measure of exactly what it says it is: the amount of energy used to power life when we are doing nothing in particular. This is the amount of energy our bodies need to keep us alive and well without any movement as such. The AEE is what the body uses to move. When you divide the amount of energy that obese and trim people use into these categories there are some interesting results. In certain groups of people, REE is high. One study showed that the obese burn one-third more energy while resting than those in the healthy weight zone; they also demonstrated a higher AEE.[25] The subjects of the study were the morbidly obese: 30 women on the waiting list for gastric bypass surgery at Tufts-New England Medical Center Hospital, Boston, USA. The heavier they were, the higher the REE, the higher the AEE and the more energy was transformed to heat after a meal. In addition, levels of both insulin and glucose in the blood were raised as body weight increased, demonstrating that there was an increasing change in body chemistry as fat accumulated.

Faster metabolism in a bigger body is hardly surprising: big people need more muscle to carry their extra padding around with them, and even resting muscle burns up calories. Additionally, raised metabolism, particularly when the energy is transformed to release heat, could be the body's way of fighting obesity: obesity is not a natural state and there is growing evidence to suggest that mechanisms are employed to prevent too much weight gain. After all, according to evolutionary theory our purpose is to procreate, and the chance of that is significantly reduced as unhealthy fat accumulation causes infertility; that the body has protection mechanisms against becoming too heavy makes perfect sense. The raised metabolic rate found in this group of obese women from Boston could be a way of burning up extra calories rather than having them stored as fat; perhaps this is an extravagance only allowed when times are so plentiful that more than adequate reserves have been made. Or it may simply be a side-effect of the stronger muscles needed to support the extra bulk. The findings show that some obese people are far from thrifty, but they are also not energy-extravagant enough to be protected against unhealthy weight gain.

Overall, the morbidly obese of Boston burn more energy, but when the metabolic rate is correlated with body mass there is a different conclusion: as body weight increases, the REE per kilogram decreases. Again, such an observation is hardly surprising because the composition of body weight is changing: the obese have a higher proportion of energy-static fat compared to energy-consuming muscle, therefore when fuel consumption is averaged over all the tissues it is bound to fall. Yet the plot is a little more convoluted: it seems that in some obese people their muscles use less fuel to move the same load when compared with the trim. In addition to the REE per kilogram falling, when the AEE is corrected for body weight, it falls as fat becomes deposited. On one hand, the total energy consumption is increased, for example via energy wastage in heat production after a meal, and on the other hand, the muscles are more efficient with their reserves, a sign of premature frailty. Over in New Zealand, studies with doubly labelled water showed that the non-obese Polynesians expended much more energy in physical activity than the non-obese Europeans;[26] physical exercise was much more effective at controlling their body weight. Yet, once individuals become obese, the AEE rates of the two populations are more similar. The incidence of obesity across that country shows that Polynesians are much more susceptible to unhealthy weight gain than Europeans. Jack Wilmore, formerly of the College of Education and Human Development at Texas A&M University, is described by this college as "One of the most influential exercise physiologists in the world during his 37-year career." When he mathematically analysed weight loss over a 20-week period among participants in an intensified exercise regime he noted that the observed loss of fat was only half what would be expected if the energy balance were fair. He proposed that his subjects had worked up an appetite, and ate more to compensate for fat loss.[27] But the results could also be explained by a more efficient AEE. Such observations could explain both the origin of obesity and its maintenance: slothfulness may reduce the capacity of the raised AEE to protect, and once fat begins to accumulate, the reduced AEE makes body weight control much more difficult. Such studies demonstrate the importance of a healthy level of exercise in weight control; they also suggest that it is easier to prevent unhealthy weight gain than to lose it later.

A more efficient AEE among the obese may seem contradictory to the observation that obesity is associated with inefficient autophagy, reducing the ability of the mitochondria to transfer the energy in acetyl units bound to co-enzyme-A into the universal fuel ATP, but in fact the observations match. When mitochondrial function falls, the capacity of the citric acid cycle to transfer energy from acetyl-co-enzyme A to ATP is reduced. The amount of NADH falls relative to acetyl-co-enzyme A, and the epigenome alters to promote body fat deposition at the expense of powering the cell. The cell is provided with less fuel and so must implement energy economy, irrespective of whether fat reserves are being deposited.

8.3 STRONG BONES

Given the first law of thermodynamics, it may seem illogical to suggest that the obese can power the same workload on less fuel, but the obese can use less to achieve more without becoming exempt from the laws that govern the universe if the efficiency of the process is changed. Transferring energy from one form to another is never complete: in addition to the desired goal, small amounts of energy are transformed to other forms, often heat, light and sound. Consider movement energy in our bodies. When we exercise we get warm because the transformation of chemical energy trapped in food to muscular movement is inefficient and produces heat. Changing the efficiency of the transformation process allows more movement from the same amount of energy. In the case of biological systems, there are many economies to be made because energy is not just allocated to the power process itself, but also to adaptation processes that preview future events. In humans, the adaptation processes in response to exercise include remodelling bones.

Bone building is regulated by the combined contribution of two signals: mechanical and chemical. The mechanical signal depends on the strain and load the bones experience, thickening – or weakening – areas according to needs; the chemical signals are more systematic, determining bone turnover according to diet, lifestyle and health – the general factors that affect body chemistry.

When we walk, each step puts pressure on different parts of the bones and the body responds by thickening the bits stressed most.

So significant is the role of exercise in shaping our bones that a trained eye can identify the arm tennis players predominantly use from X-ray images;[28] those guiding the racket are much thicker. Many aspects of our lifestyle are recorded in our bones. At the end of the 19th century, Julius Wolff defined a law governing these changes from his studies on the dead (see It all Began with Change: Shattering the Stone Age Dream).[29] He proposed an idea that has been accepted as a law – an idea that is yet to be disproven – describing how bones are remodelled by physical activity or inactivity. However, the last word of the law is as important as the others: inactivity. Inactivity can lead to bone weakening.[30]

Christopher Ruff is an archaeologist-come-detective who analyses our ancestors' bones to piece together a picture of their lives. One of his many subjects was the Ice Man Ötzi (see It all Began with Change: Taking Control). It was Ruff who suggested that his lifestyle involved walking over hilly terrain. In addition to unravelling the past, Ruff's work flashes forward to the future. In 2005, he stated "there has been a decline in overall skeletal strength relative to body size over the course of human evolution that has become progressively steeper in recent millennia, probably due to inactivity and technological advancement."[31] The long line of historical data recorded in human bones has enabled scientists to define some markers for health, and these markers show that our current diet and lifestyle are not ideal[32] – with or without obesity.

Obesity is one symptom of sequential changes in diet and lifestyle that have cumulated to undermine our health; there are other symptoms and other adverse health effects too. Ruff's statement, quoted above, does not just apply across the population, but across individual lifetimes: as unhealthy weight accumulates, the relative thickness of bones decreases. In many cases the bones of the obese are thicker than those of the trim because of the extra load they experience in lifting the fat, but they are not thickened in proportion to that load. The reduced bone deposition may be because there is an upper limit for bone strength and when this threshold is crossed there is no extra thickening. It may also be because of the type of loading experienced. Professional gymnasts have high bone density because many of the jumps they perform load the bones with five times their body weight. This type of force – jolting – is the most powerful for triggering bone strengthening. It is jolting that allows tennis players' racket arms to strengthen noticeably.

Yet heavy people are often slow-moving, loading their bones gently and so experiencing less of the most powerful bone-strengthening signals. However, there are other explanations: elevated cortisol – as a consequence of lifestyle or diet – signals for bone weakening, and the chemical signals for bone strengthening are linked to nutrition.

The dietary contributors towards bone strength that we know of are vitamin D and calcium. Many foods are fortified with calcium and vitamin D to prevent rickets. Yet all this fortification will not compensate for an imbalanced diet because the amount of nutrition absorbed from food and retained by the body strongly depends on other factors. For example, another common ingredient of supplementation is vitamin C. This anti-oxidant modifies minerals, impeding their absorption in the gut. It is not alone in its effect. Other anti-oxidants similarly modify minerals. Phytate, a naturally occurring chemical found in bran, blocks mineral absorption from a different angle: it chelates metal ions before the body can sequester them for itself and, because it is not digested, passes out of the gut with the minerals in tow. Phytate is deactivated by cooking, and so using wholemeal flours in breads leaves the minerals accessible; sprinkling bran on foods to add a fibre supplement does not. Vitamin C and bran are rightly considered an essential part of a healthy diet. These components are selling points for food products, particularly selected by people trying to improve their nutrition – perhaps trying to control their body weight via a healthy diet. Both of these food components are considered to be so beneficial there is no such thing as an excess – could Paracelsus be wrong? Yet although these components have few direct health effects, their indirect effect on mineral absorption defines an upper limit to a healthy intake, and too much could cause weight gain. The disproportionate thickening of bones seen in the obese could reflect calcium deficiencies caused or exacerbated by heavy reliance on vitamin C supplements or the wrong type of fibre.

Calcium leaching from the bones can also be triggered by other aspects of the diet, particularly excess phosphate, which has become a common additive to processed foods and drinks, and glutamate, which, as MSG, is a common ingredient used to improve the flavour of processed foods. Like vitamin C and phytate, phosphate and glutamate are essential nutrients, but their effect is not mediated at the level of nutrient absorption; rather it involves body signalling. Both phosphate and glutamate regulate the balance of calcium between the bones and the blood.

Glutamate acts as a signalling molecule that encourages the removal of calcium from teeth and bones. MSG can be used to promote bone density loss in rats; at the same time, it causes obesity and a blood lipid profile much like that increasingly observed amongst the obese.[33] The doses needed to achieve these effects are high and not likely to be achieved in the average human diet, but coupled to sedentary lifestyles, calcium deficiencies and excess phosphate, the amounts of dietary MSG some people consume may be enough to mimic the observations from the laboratory.

Since Wolff proposed his law much more information has been uncovered with respect to bones, including how bone strength alters through life. As children, we deposit bone tissue and this continues through our teens. Indeed, activity during the adolescent growth spurt is key to determining the size and shape of adult bones; certainly more than exercise later in life. Somewhere in our early 20s deposition stops, and maintaining what we have becomes an uphill struggle, but the most significant turnaround, when it comes to bone deposition, is in women after menopause. The menopause brings with it many hormonal changes, one of which signals a cut-back in bone strength, dramatically increasing the chance of fracture. The fat chemistry profile changes too. The role of bones goes beyond providing scaffolding for our soft tissues to act as a reservoir for phosphate and calcium signallers, and so altered bone strength has far-reaching consequences for health.

One hormone involved in the bone-linked control of appetite is leptin. Recall that leptin is a hunger-blocking hormone produced in a healthy body in proportion to fat deposits, and its activity is affected by diet (see Stone Age Obesity).[34] Leptin also regulates different aspects of metabolism, from energy use to bone strength, to reflect nutritional status: as fat deposits are used, leptin signalling levels fall and, with that, hunger increases and bones begin to weaken. It follows that anorexia is associated with a loss of bone mass because of a lack of leptin-producing fat. In turn, the reduced bone turnover affects metabolism, slowing it down. The effects are amplified as the disease progresses because anorexics can become resistant to leptin.[35] The molecular mechanism of this resistance is unclear, but it certainly does not generate a need to feed nor involve excess fat. It may involve the altered bone strength or nutrient deficiencies brought on by the reduced food intake. However, a lack of food is not the only cause of a lack of nutrition: a nutritionally poor diet can cause deficiencies despite providing ample energy; excess of one nutrient can cause a relative or actual deficiency of another. For example, the amounts of vitamin C and phytate eaten could cause mineral deficiencies in a relatively rich diet. Mineral deficiencies affecting bone strength may be responsible for leptin deficiency among the obese.

Leptin's power is not just linked to its production; it is also linked to its ability to transmit its signal. Leptin resistance is usually associated with obesity: it is one of the potential causative changes in fat chemistry that seems to be common among the obese. In the anorexic state, one could imagine that leptin resistance would step up hunger to encourage the patient to eat – if possible. Among the obese, leptin resistance simply stocks fat. But in both cases – obesity and anorexia – leptin resistance reduces bone strength. Bone turnover can be monitored by a simple blood test to measure the amount of a molecular marker called osteocalcin: the higher the level of osteocalcin in the blood the higher the level of bone turnover. Steve Bloom, the head of the Division of Investigative Science at Imperial College London and clinical director of pathology at Hammersmith Hospital, has shown that both leptin-deficient and leptin-resistant mice have higher levels of this marker in their blood; the bones are weakening.[36] Amongst humans, there is a similar effect: the mechanical stress put on bone caused by carrying the burden of obesity usually ensures that bones stay strong enough to resist fracture, but not as strong as they should be if leptin signalling was working correctly.

Resistance to leptin has two consequences: first, leptin resistant people lose bone mass and, second, leptin resistance causes appetite control to be knocked out of balance with the body's needs, leading to overeating. These combined effects question the role of this signaller in anorexia until we consider the work of Gerard Karsenty, Professor and Chair of Genetics and Development at the Columbia University Medical Center. He demonstrated that the two effects were mediated by different mechanisms, chemically blocking leptin's role in appetite regulation leading to body weight gain, without affecting its role in regulating bone density.[37] The chemical he used to achieve this effect was MSG. The amounts he used were high; a poisoning dose. But the work still raises the question whether this food additive could be contributing to the obesity problem. It is unlikely to cause the problem alone, but combined with phosphate, salt and aspartate excesses and calcium and vitamin D deficiencies, the case becomes stronger. Leptin is the body's fat-stat, a key signaller in appetite regulation. Could it be that compensatory appetite regulation is dependent on diet and, if this is the case, could it be that the reason fat teenagers are unable to compensate for a fast-food calorie boost is that their diet exceeds the tolerance limits of MSG?

There are other reasons for leptin resistance. For example, triacylglycerides (TAGs) block the signalling capacity of this hormone. When leptin was discovered it was dubbed a fat-stat because its first identified function was communicating the state of fat stores to the brain. The message has a complex journey because of the blood–brain barrier. Leptin must hook up with a special carrier that allows it to transmit its signal across this wall. Without the carrier the signal goes unheard. TAGs block the leptin–carrier interaction. The role of blood fat in blocking leptin's signal may seem counter-intuitive because when the body is well fed one would imagine that appetite would be quenched, but, as with galanin signalling, the reverse is true: the more fat we eat, the hungrier we become. Fat and sugar in the diet can increase blood TAG levels. These food components tend to be higher in diets based around processed foods. Leptin signalling can also be rekindled by dietary fibre. This food component tends to be lower in a processed food diet. Fibre is important for digestive health, but we can survive, at least in the short term, without it. It is neither a macronutrient nor a micronutrient because it does not provide nutrition. However it does aid digestion, stimulate the production of appetite-suppressing hormones, and promote insulin sensitivity; it has health benefits. And, even with the effect of phytate on mineral absorption, the amount of fibre in the diet is inversely correlated to BMI.[38]

High-fibre diets improve insulin sensitivity by increasing the amount of a signaller called adiponectin in the blood;[39] this signals for more fat burning in the muscles, healthy fat storage and a faster metabolic rate. As adiponectin levels increase, more fuel is taken from the blood into cells. The fuel is mainly leptin-blocking TAGs; thus, adiponectin promotes fat metabolism and, by mopping up these leptin-blockers, helps appetite suppression. Thereby adiponectin production is central to maintaining a healthy body weight.

Like leptin, adiponectin is produced in the adipose tissue around the body, so it would be expected that the more fat stored, the more this fat-burning signaller is produced and the more fat is used as the favoured fuel, preventing unhealthy weight gain. However, despite the fact that obese people have more fat, they often produce less of this signaller.[40] Quite what causes this decrease in production is difficult to say, but it may be simply due to a lack of dietary fibre.

Therefore, the type of menu that hits bone strength, metabolism and appetite with the heaviest consequences would be low in fibre, lacking calcium and vitamin D, high in fat and sugar, rich in MSG, perhaps with a splash of aspartame, and washed down with some dilute phosphoric acid; it fits the description of a burger, fries and cola.

8.4 NO SET-POINT

Appetite control combines with the metabolic rate to determine body fat deposition. In 1982, US nutritionists William Bennett and Joel Gurin proposed that the body predetermined a level of body fat to store and strove to ensure that reserves remained at that level, irrespective of whether it was healthy or not.[41] They called this idea the set-point theory.

The set-point theory is based on the assumption that the body has an internal register – a fat-stat that regulates the size of energy stores. Just as a thermostat can be adjusted to regulate temperature at a pre-selected point, the fat-stat can be adjusted to regulate body weight at any given value: healthy, a little too light, a little too heavy, or fatal. When we overeat and the fat-stat relays the message that stores are full, compensatory appetite regulation kicks in and reduces our desire for food. And when fat stores are tapped to make up for a shortfall in food intake, the fat-stat registers the falling reserve and we feel hungry. The proposed appetite regulation is based on what the body thinks our body weight should be – the set-point.

When the set-point theory was presented, it was readily embraced by the public. To the layman, it explained the problems people experience with weight loss; particularly, why weight loss is often followed by re-gain; why sometimes dieters reach a plateau where they seem to be unable to shed pounds; and why some people are seemingly resistant to obesity. The scientific domain was less receptive. There was some evidence to support the theory, for example the observation that over a five-year period in the 1980s, the body weight of a surveyed section of the Dutch population did not change despite dietary improvement.[42] But there was insufficient evidence to lift the natural scepticism scientists have for their peers' work; the case of the Dutch could be explained away by increasingly sedentary lifestyles.

The set-point theory fits some data, but conflicts with other observations. For example, the theory explains why 1950s Britons did not waste away despite eating less than the current calorie recommendations: energy economy may have been triggered because their set-point was in the healthy weight range. It also explains how the Lithuanians manage to maintain a trim physique despite apparently eating much more than they need: energy extravagance is triggered because their set-point is in the healthy weight range. It explains why lean teenagers have compensatory appetite regulation mechanisms that mean that, if they feast one day, they eat less the next: their appetite regulation suggests that they have a set-point in the healthy weight range. In contrast, their obese peers do not share the same feasting profile; either their set-point is elevated beyond their current body weight or they have no set point at all. Similarly, the Maltese seem to have a higher than healthy set-point. However, for widespread scientific acceptance there needs to be a molecular mechanism to explain how the set-point is set and maintained and to explain how it can be lost.

Bennett and Gurin suggested that the only way of lowering the predetermined set-point value is through cigarette smoking or exercise. The incidence of smoking has decreased with the rise of the obesity epidemic, but it is worth remembering that before smoking was introduced to the West, there was no epidemic of obesity, and in some countries of the world, such as India, there is an obesity epidemic despite widespread tobacco use. In contrast, sedentary lifestyles are new. Leptin is a fat-stat that communicates the level of fat stores to the brain. Exercise increases leptin signalling, both via bone strength and burning TAGS, hence Bennett and Gurin's exercise link could be explained by leptin signalling.

The amount of energy consumed by exercise (the AEE) has been suggested to be a factor in susceptibility to obesity (see A Fat Future? The Memory of Malnutrition). As body fat rises the AEE seems to fall, suggesting that energy conservation mechanisms are in place. One possible mechanism to explain this observation is via the TAG–leptin interaction. TAG levels are characteristically raised amongst the obese, and leptin-resistance is also known. Given that exercise consumes TAGs, it allows fat reserves to be more accurately gauged via the leptin signal, giving a role for exercise in health beyond consuming calories. When Bennett and Gurin observed that exercise reduced the body weight set-point, what they actually may have seen was leptin signalling being restored to promote healthy weight control. When exercise fails to work its health-restoring magic, it may be because the TAG levels have been raised so significantly and the AEE rate has fallen so low that the level of activity required to restore healthy signalling has gone beyond that easily attainable.

Leptin has now been shown to have regulatory roles beyond a fat-stat, but this first identified function could be behind the obesity epidemic. The set-point theory assumes a mechanism to monitor the amount of stored fat and leptin fills the role. But some people – the obese – seem to have no set-point and this effect could also be explained by what we know of leptin, or rather resistance to this signaller. Regardless of whether the mechanism of resistance is via altered bone strength caused by sedentary lifestyles or dietary factors – MSG, phosphate, vitamin C, calcium, phytate, TAGs, or their sweet precursor, to name but a few potential factors – it is possible to estimate the power of the effect because leptin-null genes are models of complete leptin resistance and the result is morbid obesity due to an overwhelming appetite. When there is no leptin signalling, there is no body weight set-point. On the other hand, there are variations in the leptin gene that cause variety in the population within the healthy weight zone; hence, it is likely that the body can compensate for low levels of resistance, but the capacity for compensation is likely to be strongly affected by diet. As an illustration, consider the case of two groups of slightly leptin-resistant rodents, one fed a fatty diet and the other the lean equivalent: after just two months, the high-fat fed group gained 84% more weight than those with a similar leptin profile fed on a low-fat diet. The unhealthy body fat was disproportionately deposited on the abdomen.[43] Clearly the leptin resistance alone did not cause the unhealthy weight gain, because the rodents fed on the low-fat diet did not gain unhealthy body weight, but the leptin resistance abolished the ability to stay trim on a high-fat diet.

At first the leptin-resistant rodents fought back against the fatty food, making adaptations such as increasing the number of leptin receptors to amplify its message, and raising the REE. They tried to maintain their set-point. But after the high-fat diet had been sustained over several months, the compensation ceased and not only did the initial boost of leptin sensitivity drop, it dropped below the starting value.[43] Effectively, their set-point was raised because more leptin was needed, equating to more leptin-producing body fat, to satisfy hunger. A high-fat diet raised the set-point.

8.5   A BURDEN ON THE POOR

Of all states in the USA, Mississippi has the highest incidence of obesity; it also has the lowest income per capita and the lowest standard of health care. This same wealth–weight relationship is common within and between the industrialised countries of the world: often those people that have the most access to food are the slimmest and those that have least tend to be the heaviest.

France reports wealth–weight relationships using statistics: in Lille, over a 10-year period from 1989, the number of obese five year olds more than doubled.[44] The big rise in weight was among those children from the lowest-income families. In fact, the children from the highest social classes showed no change in the average amount of fat stored, but in the lowest-income bracket body weight soared: both the number of obese increased and the average size of their bodies grew. Interestingly, in the middle-income bracket, the heavy became heavier without the number of obesity cases changing; the set-point for the already heavy was raised, but the trim retained their protection. These changes support the idea that obesity is a self-perpetuating disease; the greater the amount of fat stored in excess of the healthy level, the easier it is to add more.

Poverty promotes obesity at all stages of life. People in their late 30s are notoriously prone to middle-age spread and it has long been accepted that body weight increases are more common as we grow older, possibly due to the accumulation of oxidative damage. But how rapidly we spread depends on wealth: once obese, the lower the income the faster the body weight tends to accumulate in midlife.[45] About 20 000 men and women aged between 35 and 55 years volunteered for the Stockport population-based cardio-vascular disease risk factor screening programme, which included collecting information about BMI. Across the board, regardless of weight, age and income, the average BMI rose during the four-year period analysed, but among the obese, a lower income significantly increased the chance of weight gain. These results suggest that there are differences in lifestyle and diet between different British socioeconomic groups in place before the age of 35, promoting obesity in the lower-income groups and offering some protection to the wealthy. Yet the studies were not detailed enough to elucidate the nature of wealth-related obesity resistance: it could be education with respect to exercise or diet; stress levels affecting cortisol production; malnutrition mediating its effect directly via a signaller or indirectly altering the way we use genes; or it may be a lifestyle-linked effect with a lack of exercise – or even sunshine – altering fat chemistry to promote unhealthy weight gain. The changes may be transient or long term. Irrespective, all these causes are curable and avoidable.

There is a north–south divide in the incidence of obesity in the UK: Britons living in the north tend to be plumper than those living in the south. The average income per capita is higher in the south than the north. Similarly, a north–south obesity divide is found in Italy, but in reverse; southern Italy has one of the highest incidences of childhood obesity in Europe, with just over one-third of under-fives weighing in as obese – and it is one of the poorest regions of Western Europe too. The wealthier northern Italians are more likely to be lean than their southern compatriots.[46] The reverse relationship in these comparisons excludes the cold as a driving force towards obesity; it does, however, support the idea that the risk of becoming obese is inversely correlated to wealth.

Austria is one of the wealthiest countries of the world, but there is a significant distribution of wealth within its land-locked boarders. Even in a country as small as Austria, there are regional differences in the tendency to store fat, with those living in the west being more resistant to obesity than those in the east,[47] with one exception: Vienna. The country's capital is nestled in the east among the flat farming plains. Within the cosmopolitan capital people are trim and trendy, but venture outside the city limits and it is a different story: the incidence of obesity among young farmers doubled over the two decades leading up to 2005.[48] Farming is physically hard work, which would be expected to keep obesity at bay – but farming has generally poor financial returns.

For the Belgians, a Flemish background slightly reduces the likelihood of becoming obese compared with the French alternative. These peoples share the same locales, excluding climate and environmental pollutants as causes of obesity. Genes may contribute to the body fat distribution as the Belgians were divided by their ancestry. And there are indications of a behavioural difference: the Belgian-Flemish eat more vegetables than the Belgian-French, but fewer fruits and sweets.[49] The Belgian-Flemish are also arguably the wealthier of the two sectors.

Amongst UK teenagers, eating habits have been linked to adult body weight, and those with a lower income tend to have the worst eating habits and the highest BMI. At the beginning of this millennium, the UK National Diet and Nutrition Survey questioned nearly 2000 adults aged 19 to 64 years about food habits and nutrition. It uncovered evidence to suggest that the diets of younger adults, particularly women and those in lower socioeconomic groups, lacked several important nutrients.[50] The consequences of teenage malnutrition are three-fold: they could trigger metabolic changes affecting current health; they could also increase the susceptibility to obesity later on in life; and these nutritional problems could increase the chance of obesity in the next generation. One nutrient identified to be deficient among these adults was vitamin D. Studies show that such deficiencies promote obesity, hypertension, insulin resistance and progression to diabetes mellitus. The vitamin D requirement among the obese is elevated because this nutrient is fat soluble, migrating into the adipose tissue and becoming trapped there. This nutrient is used in the fortification of foods, including dairy products, but this enrichment does not seem to meet the nutritional needs of these teenagers; perhaps it cannot balance the nutritionally poor diet. And they are not alone: over 200 obese children who received a standard physical examination at the paediatric endocrine clinic of the Infants and Children's Hospital of Brooklyn at Maimonides, New York, were assessed for vitamin D levels and over half were deficient.[51]

When 264 Bulgarian factory workers filled in a questionnaire, the results matched what we would expect from our molecular knowledge of fat chemistry. Over half of the people were overweight and about the same number were eating more than the recommended daily intake of protein.[52] Locked in Liebig's ideas that dietary protein is good, the questions Moura and others have raised about how much protein is too much have yet to have an impact on the public's understanding of nutrition. The factory workers also ate a calorie surplus: a simple observation that shows that their appetite control with respect to calories was not functioning as it should. Given that part of the calorie surplus was in fat, one could speculate that galanin was at work stimulating the appetite for fatty foods in response to these same foods in the diet. The fats in the blood could also fuel hunger by blocking leptin signalling, a double whammy against appetite control. The survey suggested that these people also ate too much salt, affecting both their sodium balance and glutamate clearance – and body weight regulation. However, in their diet of plenty, they failed to meet the recommended daily dose of hunger-busting fibre and vitamins important for energy metabolism. The factory workers had a body-weight promoting diet; they also had a diet that promoted ageing and other degenerative diseases.

The 1997 Egyptian Integrated Household Survey correlated micronutrient deficiencies with being overweight.[53] Abay Asfaw of the International Food Policy Research Institute based in the USA looked carefully at these data and made two important observations. First, he asserted that the impact of micronutrient deficiencies on health is not well known; certainly the understanding is not deep enough to allow complacency with respect to nutrition. Second, he concluded that the Egyptian food subsidy programme put in place to help nourish low-income families is not meeting the needs of the people it is designed to help. This programme, he states, "lowers the relative prices of energy-dense, nutrient-poor food items, can be one of the major factors for the emergence of overweight/obese and micronutrient deficient mothers in the country." Isn't this policy the same the world over? Could this be why the poor are more vulnerable to obesity than the rich?

Barbara Rolls of Pennsylvania State University, USA, showed that women who were served a low-calorie salad before helping themselves to pasta ate about 12% less than those who had skipped the starter.[54] Worryingly though, women who were served a starter containing nearly a quarter of their recommended daily calorie allowance in fat ate more than women who had no starter at all. These observations are evidence to suggest that appetite control is not based on calories alone, nor on portion size alone, but is reflected in the types of food eaten – the quality of the food. A green salad starter quenches appetite without too many additional calories, but opt for deep-fried onion rings and potato skins and the opposite effect is observed.

Outside the laboratory environment, the effect of salad starters on appetite can be loosely gauged by comparing the French and the British because, although a *un peu de salade* is often included in the price of a meal in France, many UK families do not have the routine of starting the main meal with raw veg. The Britons are just as likely to opt for deep-fried alternatives or skip the starter in favour of a chocolate pudding. These high-calorie foods stimulate rather than suppress the appetite. Mouthful-for-mouthful, high-carb or high-fat foods are generally cheaper than fresh salads and vegetables. And if poverty pushes appetite-quenching food off the menu, filling the spaces with low-cost, low-nutrient alternatives is a certain route to weight gain.

A poverty-linked, low-nutrient diet explains the distribution of obesity within countries and also between many of the states of the USA, but between countries it is not so straightforward. Latvians and Lithuanians seem to have obesity resistance. Over the first decade of this millennium, their gross domestic product (GDP) per capita was approaching one-quarter of that of the USA; yet the wealthy North Americans have more of an obesity problem. But dig a little deeper and poor nutrition comes back as a possible explanation: the GDP per capita does not always reflect the quality of the nutrition in the country.

In some countries of the world the obesity epidemic has hit so hard and fast that records are available from before fat became a problem through to when it became the number one cause of premature death. In the majority of cases, the dietary changes during the transition include the increased consumption of high-calorie, low-nutrient processed foods; many of these same foods are marketed in the USA; many of these same foods have a chemical balance that promotes body fat.

The influence of British colonial rule over Malta for a period of more than 150 years and the impact of tourism in the latter half of the 20th century have precipitated dietary traditions more in line with those of English-speaking nations than the geographically closer olive-and-vegetable based Mediterranean diet.[55] The typical Med diet is not only a matter of health but a matter of pride, with families boasting of the quality of their olive oils and size of their home grown veg, but in Malta population density makes home grown food a luxury and for the majority of the population it is unfeasible. Therefore, the country relies heavily on food imports: Cheddar cheese, sugar, tinned meats and condensed milk are some of the favourites. Over the last four decades, the Maltese have increased their food consumption all round; for example, between 1961 and 1991 consumption of cereal products and fish stayed about the same, but the consumption of meat – processed or otherwise – and fruits and vegetables combined both rose threefold – healthy changes and not so healthy changes. It is not clear which change has triggered the wave of obesity, but it is clear that among the Maltese body chemistry is no longer able to run at an automated level to protect against unhealthy weight gain. And the Maltese are not alone in their blight.

Jump over the ocean to the Pacific Islands and similar transitions have taken place. Back in the 18th century, when the culture was unpolluted, food had symbolic and economic importance and excessive body weight was a sign of wealth. Yet despite the lure of being big, most people were described as being "strong, muscular and mostly in good health",[56] and they remained that way until World War Two. After this global upheaval, the islands of Oceania were either colonised or protected by various industrialised nations of the world: the USA, Australia, New Zealand, France or the UK. And the protection includes food aid. Like the Maltese, when the Pacific Islanders were introduced to fast food, white rice and spam the traditional foods became a little less popular and obesity became the norm;[57–59] the aid eliminated the need for domestic fishing and farming. Now 75% or more of adults are reported as obese in Nauru, Samoa, American Samoa, the Cook Islands, Tonga and French Polynesia.[60] Just as in China, unhealthy body weight is prevalent in urban areas and those that maintain a more traditional lifestyle also maintain their health.[61]

In the West, fast foods are often the cheapest stomach fillers we can buy in terms of money for calories. However, those who, on average, have a fast-food stop more than twice a week gain 300 g of extra body weight each year compared with those who rarely touch this type of food.[62] Over a decade and a half, fast-food eaters gain an average of 4.5 kg. In many cases the weight gain is unwanted, but it is also unhealthy: these people double their chance of developing diabetes. Fast food and convenience foods are the ultimate in meeting energy needs – with little consideration of the body's nutritional needs. The consumption of these products could explain the cultural factor that protects against obesity. Consider the market for fast food in the USA, UK and France. The US market is 10 times that of the UK, and the UK market is three times that of France. Many of these products are marketed as franchises and countries with a low GDP per capita, such as Latvia and Lithuania, have only recently been able to buy a stake. There is a correlation between fast-food consumption and the incidence of obesity in many countries. In a typical day in the USA about one-third of children will eat fast food rather than a balanced meal.

The rise in the incidence of obesity with such dietary change demonstrates that there is something about energy-dense nosh that causes us to gain weight. It could be the result of calorie availability coupled to greed, but the Latvians and French are living demonstrations to disprove that idea! It could be the combination of fats and sugars blocking leptin signalling. It could be a lack of antioxidants in the diet; such nutrients are often found in fresh produce. It could also be the result of nutrient excess: vitamin C supplements or phytate preventing calcium absorption. It could be that lifestyle has become too sedentary to support enough bone strength to regulate metabolism, or too sedentary to burn the minimum energy intake necessary to ensure a balance of other nutrients. It could also be combinations of additives or deficiencies of nutrients that alter body chemistry to promote weight gain. Even with imports of Western foods preventing hunger on the Pacific Islands, malnutrition is rife: anaemia, B vitamin deficiency and calcium deficiencies are typical amongst these peoples.

Budget staples are part of a worldwide initiative to provide enough food for the population and quench some formerly devastating types of malnutrition, but with a focus on calories and a handful of nutrients, a diet of such staples is not always balanced. Traditionally, malnutrition was a result of insufficient food in general, but now the cheap, highly processed foods cause a new wave of nutritional deficiencies and excess.

Switzerland boasts one of the lowest obesity statistics in Europe, but their resistance is being challenged. Over the past decade there has been rapid change in the supermarkets and in the style of eating out, and the incidence of unhealthy body weight has started to rise in a way that suggests there is an epidemic on the horizon. The average age at the onset of obesity is falling and there is an increasing incidence of the disease amongst children. The changes are in part due to immigration, because some people arrive with excess body baggage. It is also a result of cultural change: expats demand their own food favourites, including commercial chain coffee shops, fast-food restaurants and convenience foods. More choice has arrived in the supermarkets. The expanded range of products includes a budget range as well as flavours from around the world. Switzerland has imported new international foods, including flavoured potato snacks, processed chicken cuts and a whole range of ready meals – a new international lifestyle and a new international body shape.

The problem with nutritionally poor meals is not just the extra fat, salt and sugar they contain, but the fact that these foods usually replace something more nutritious in the diet. To put these nutritional differences into numbers, a study of over 6000 children aged between four and nineteen years compared the nutrition of those children who ate fast food once with those who skipped such stomach fillers for a day.[63] Opting for fast food resulted in the children consuming an excess of 10% of their recommended daily calorie allowance, 9 g more fat, 24 g more carbohydrate and 26 g more added sugars. In addition, these children consumed 1 g less fibre, 65 g less milk and 45 g fewer fruits and non-starchy vegetables – these differences are in just one day and just one fast-food meal. Many children have fast food or fast-food style school lunches every single day. It is worth reiterating what these dietary changes mean: more calories promote weight gain, even if the link is not directly proportional; more fat and sugar blocks leptin, exacerbating the contribution of the calories; more sugar could cause insulin spikes promoting fat storage rather than burning; both the fat and sugar increase galanin production, increasing appetite; less fibre means less hormonal appetite suppression and less adiponectin to aid leptin signalling, thereby increasing appetite; less milk means less calcium, and if the milk is replaced by mineral-striping cola, the effect is both bone weakening and body weight gain; fewer fruits and vegetables mean less fibre and fewer vitamins, including phytochemicals that protect against degenerative diseases, cancer and bacteria leaving the fight solely to the immune system; fewer anti-oxidants increases the chance of histone acetylation, altering gene use to promote fat storage. No doubt the fast food contains plenty of salt, to harden the pulse and promote body weight gain, and a dash of MSG added – perhaps some artificial sweetener too – to interfere with the glutamate signalling pathway. This type of dietary change does not just change the outward form, but deprives the body of what it needs to achieve a full and healthy life. And when obesity begins in childhood health declines rapidly – not just in that generation but, via epigenetics, in the next generation too.

REFERENCES

1. M. Duclos, J. B. Corcuff, N. Etcheverry, M. Rashedi, A. Tabarin and P. Roger, *J. Endocrinol. Invest.*, 1999, **22**, 465–471.
2. T. F. Oberlander, J. Weinberg, M. Papsdorf, R. Grunau, S. Misri and A. M. Devlin, *Epigenetics*, 2008, **3**, 97–106.
3. R. Z. Levene, B. Schwartz and P. L. Workman, *Arch. Ophthalmol.*, 1972, **87**, 389–391.
4. A. W. Meikle, J. D. Stringham, M. G. Woodward and D. T. Bishop, *Metab. Clin. Exp.*, 1988, **37**, 514–517.
5. M. Bartels, M. Van den Berg, F. Sluyter, D. I. Boomsma and E. J. C. de Geus, *Psychoneuroendocrinol.*, 2003, **28**, 121–137.
6. G. Chandramohan, N. Durham, S. Sinha, K. Norris and N. D. Vaziri, *Metab. Clin. Exp.*, 2009, **58**, 1424–1429.
7. G. Chandramohan, R. S. Diego and H. Ward, *Am. J. Hypertens.*, 2001, **14**, 93A.
8. M. Humphreys, *Am. J. Physiol. Regul. Integr. Comp. Physiol.*, 2004, **286**, R417–R430.
9. J. van Ree, B. Bohus, K. Csontos, W. Gispen, H. Greven, F. Nijkamp, F. Opmeer, G. de Rotte, T. van Wimersma Greidanus, A. Witter and D. de Wied, *Ciba Found. Symp.*, 1981, **81**, 263–276.
10. L.-I. Larsson, *Lancet*, 1977, **310**, 1321–1323.
11. H. Biebermann, T. R. Castaneda, F. van Landeghem, A. von Deimling, F. Escher, G. Brabant, J. Hebebrand, A. Hinney, M. H. Tschop, A. Gruters and H. Krude, *Cell Metab.*, 2006, **3**, 141–146.
12. G. Tonnarini, G. Dellefave, M. Chianelli, P. Mariani and M. Negri, *Eur. J. Gastroenterol. Hepatol.*, 1995, **7**, 357–360.
13. J. S. Han, *Neurosci. Lett.*, 2004, **361**, 258–261.

14. Y. Agrawal, *Med. Hypotheses*, 2005, **64**, 721–724.
15. http://clinicaltrials.gov/ct2/show/NCT00663117 (accessed 8 June 2010).
16. A. K. Traore, O. Thiero, S. Dao, F. F. C. Kounde, O. Faye, M. Cisse, J. B. McCandless, J. M. Zimmerman, K. Coulibaly, A. Diarra, M. S. Keita, S. Diallo, I. G. Traore and O. Koita, *J. AIDS HIV Res.*, 2011, **3**, 189–198.
17. L. R. Webster, *Expert Opin. Investig. Drugs*, 2007, **16**, 1277–1283.
18. H. J. Lenz and M. R. Brown, *Brain Res.*, 1987, **413**, 1–9.
19. A. Kjellgren, U. Sundequist, U. Sundholm, T. Norlander and T. Archer, *Soc. Behav. Pers.*, 2004, **32**, 103–115.
20. L. P. Grillo, A. F. A. Siqueira, A. C. Silva, P. A. Martins, I. T. N. Verreschi and A. L. Sawaya, *Eur. J. Clin. Nutr.*, 2005, **59**, 835–842.
21. E. Ravussin, S. Lillioja, W. C. Knowler, L. Christin, D. Freymond, W. G. H. Abbott, V. Boyce, B. V. Howard and C. Bogardus, *N. Engl. J. Med.*, 1988, **319**, 518–519.
22. R. Uauy and J. Kain, *Pub. Health Nutr.*, 2002, **5**, 223–229.
23. J. Gonzalez-Barranco and J. M. Rios-Torres, *Nutr. Rev.*, 2004, **62**, S134–S139.
24. L. R. Dugas, R. Harders, S. Merrill, K. Ebersole, D. A. Shoham, E. C. Rush, F. K. Assah, T. Forrester, R. A. Durazo-Arvizu and A. Luke, *Am. J. Clin. Nutr.*, 2011, **93**, 427–441.
25. S. K. Das, E. Saltzman, M. A. McCrory, L. K. G. Hsu, S. A. Shikora, G. Dolnikowski, J. J. Kehayias and S. B. Roberts, *J. Nutr.*, 2004, **134**, 1412–1416.
26. E. C. Rush, L. D. Plank and W. A. Coward, *Am. J. Clin. Nutr.*, 1999, **69**, 43–48.
27. J. H. Wilmore, J. P. Despres, P. R. Stanforth, S. Mandel, T. Rice, J. Gagnon, A. S. Leon, D. C. Rao, J. S. Skinner and C. Bouchard, *Am. J. Clin. Nutr.*, 1999, **70**, 346–352.
28. H. Haapasalo, S. Kontulainen, H. Sievänen, P. Kannus, M. Järvinen and I. Vuori, *Bone*, 2000, **27**, 351–357.
29. J. Wolff, *Br. Med. J.*, 1893, **1**, 124.
30. J. H. Scott, *Am. J. Phys. Anthropol.*, 1957, **15**, 197–234.
31. C. B. Ruff, *Musculoskelet. Neuronal. Interact.*, 2005, **5**, 202–212.
32. C. Ruff, *Ann. Rev. Anthropol.*, 2002, **31**, 211–232.
33. S. Patil, T. Prakash, D. Kotresha, N. R. Rao and N. Pandy, *Ind. J. Pharmacol.*, 2011, **43**, 644–647.

34. P. G. Cammisotto and L. J. Bukowiecki, *Am. J. Physiol. Regul. Integr. Comp. Physiol.*, 2004, **287**, R1380–R1386.
35. M. Sato, N. Takeda, H. Sarui, R. Takami, K. Takami, M. Hayashi, A. Sasaki, S. Kawachi, K. Yoshino and K. Yasuda, *J. Clin. Endocrinol. Metab.*, 2001, **86**, 5273–5276.
36. S. Yasumura, J. F. Aloia, C. M. Gundberg, J. Yeh, A. N. Vaswani, K. Yuen, A. F. Lomonte, K. J. Ellis and S. H. Cohn, *J. Clin. Endocrinol. Metab.*, 1987, **64**, 681–685.
37. S. Takeda, F. Elefteriou, R. Levasseur, X. Liu, L. Zhao, K. L. Parker, D. Armstrong, P. Ducy and G. Karsenty, *Cell*, 2002, **111**, 305–317.
38. K. Hendricks, K. Wilis, R. Houser and C. Y. Jones, *J. Am. Coll. Nutr.*, 2006, **24**, 321–331.
39. L. Qi, E. Rimm, S. M. Liu, N. Rifai and F. B. Hu, *Diabetes Care*, 2005, **28**, 1022–1028.
40. M. Gil-Campos, R. Canete and A. Gil, *Clin. Nutr.*, 2004, **23**, 963–974.
41. W. Bennett and J. Gurin, *The Dieter's Dilemma: Eating Less and Weighing More*, Basic Books Inc. Publishers, New York, 1982.
42. M. R. H. Lowik, K. Hulshof, L. J. M. van der Heijden, J. H. Brussaard, J. Burema, C. Kistemaker and P. J. F. de Vries, *Int. J. Food Sci. Nutr.*, 1998, **49**, S5–S68.
43. S. Lin, L. H. Storlien and X. F. Huang, *Brain Res.*, 2000, **875**, 89–95.
44. M. Romon, A. Duhamel, N. Collinet and J. Weill, *Int. J. Obes.*, 2005, **29**, 54–59.
45. G. Lyratzopoulos, P. McElduff, R. F. Heller, M. Hanily and P. S. Lewis, *BMC Public Health*, 2005, **5**, 125–134.
46. C. Maffeis, A. Consolaro, P. Cavarzere, L. Chini, C. Banzato, A. Grezzani, D. Silvagni, G. Salzano, F. De Luca and L. Tato, *Obesity*, 2006, **14**, 765–769.
47. E. Schober, B. Rami, S. Kirchengast, T. Waldhoer and R. Sefranek, *Eur. J. Pediatr.*, 2007, **166**, 709–714.
48. T. Dorner, B. Leitner, H. Stadimann, W. Fischer, B. Neidhart, K. Lawrence, I. Kiefer, T. Rathmanner, M. Kunze and A. Rieder, *Sozial-Und Praventivmedizin*, 2004, **49**, 243–246.
49. I. Janssen, P. T. Katzmarzyk, W. F. Boyce, C. Vereecken, C. Mulvihill, C. Roberts, C. Currie and W. Pickett, *Obes. Rev.*, 2005, **6**, 123–132.

50. G. Swan, *Proc. Nutr. Soc.*, 2004, **63**, 505–512.
51. M. Smotkin-Tangorra, R. Purushothaman, A. Gupta, G. Nejati, H. Anhalt and S. Ten, *J. Pediatr. Endocrinol. Metabol.*, 2007, **20**, 817–823.
52. M. Koleva, A. Kadiiska, V. Markovska, A. Nacheva and M. Boev, *Cent. Eur. J. Public Health*, 2000, **8**, 10–13.
53. A. Asfaw, *Econ. Hum. Biol.*, 2007, **5**, 471–483.
54. B. Rolls, L. Roe and J. Meengs, *Obesity Research*, 2004, **12**, A5.
55. E. Helsing, *Food Policy*, 1991, **16**, 371–382.
56. R. t G. Hughes, *Diet, Food Supply and Obesity in the Pacific*, 2003, http://www.wpro.who.int/pdf/NUT/diet_food_supply_obesity.pdf (accessed 29 October 2009).
57. P. Zimmet, P. Taft, A. Guinea, W. Guthrie and K. Thoma, *Diabetologia*, 1977, **13**, 111–115.
58. D. Shmulewitz, S. C. Heath, M. L. Blundell, Z. H. Han, R. Sharma, J. Salit, S. B. Auerbach, S. Signorini, J. L. Breslow, M. Stoffel and J. M. Friedman, *Proc. Nat. Acad. Sci. USA*, 2006, **103**, 3502–3509.
59. E. Ravussin, P. H. Bennett, M. E. Valencia, L. O. Schulz and J. Esparza, *Diabetes Care*, 1994, **17**, 1067–1074.
60. M. Curtis, *J. Devel. Soc. Transform.*, 2004, **1**, 37–42.
61. H. Ringrose and P. Zimmet, *Am. J. Clin. Nutr.*, 1979, **32**, 1334–1341.
62. M. A. Pereira, A. I. Kartashov and C. B. Ebbeling, *Lancet*, 2005, **365**, 1030–1030.
63. M. A. Cowley, J. L. Smart, M. Rubinstein, M. G. Cordan, S. Diano, T. L. Horvath, R. D. Cone and M. J. Low, *Nature*, 2001, **411**, 480–484.

CHAPTER 9

Your Choice

If I were to draw out a map of the signalling pathways that regulate body fat deposition it would involve hundreds of molecules
and many triggers. Many molecules would respond to multiple
triggers; many triggers would affect multiple molecules. The triggers would include the calories in our foods balanced by the
energy used by the body: the energy balance. But, in addition,
the very nature of food calories would cause a characteristic
wave of signalling. Virtually calorie-free nutrients would be
shown directly tweaking appetite and metabolism: the B vitamins
to ensure efficient energy release, for example. The exercise we
take would directly burn calories, but also lower TAGs in the
blood, encourage a healthier flow of fat that protects against oxidative damage and promote body weight-regulating calcium
signalling.

And then there would be a second sphere to the map, one that
describes how nutrients interact with each other in the gut to affect
absorption and how in the body relative amounts of some nutrients
are important for health. This area is where non-nutritional signals,
such as the mechanical signals of exercise on bone strength, the
effect of mood, psychological effects and even sunshine would be
mapped.

The third sphere would show how body fat itself alters appetite
regulation and metabolism to trigger disease, including unhealthy
weight gain. A quick glance at this section would reveal the

Fat Chemistry: The Science behind Obesity
Claire S. Allardyce
© Claire S. Allardyce 2012
Published by the Royal Society of Chemistry, www.rsc.org

importance of maintaining a healthy body weight. This is where the epigenetic effects that partially determine our children's health would lie.

From this map, it would be possible to understand how the human race survived through the Stone Age with poor nutrition and energy deficiencies. Appetite regulation optimised taste preferences for survival, favouring fatty and sweet foods. Galanin ensured that when these foods were available appetite perked up. But leptin kept fat reserves out of the overweight BMI zone. It would be possible to understand how Britons from the Middle Ages through to the fabulous 50s managed to eat a fatty diet without unhealthy body weight gain – although their health suffered. One could postulate that zinc or other mineral deficiencies quenched appetite, overcoming galanin's effect, and bone strength resulting from exercise kept the metabolic rate high. In this era, there was a constant battle with unhealthy body weight, but the desire to be slim was stronger than the drive towards obesity.

In the last few decades, the power struggle has reversed. The desire to be slim is no longer greater than the drive towards obesity. Epidemics of unhealthy body fat are the result. There are many psychological effects that could be used to explain the change, but there is also a fat chemistry solution. The key factors in body weight control include elements of healthy living beyond calories: aspects of lifestyle and nutrition that affect endorphin production and cortisol balance, for example. In the case of both signallers, sedentary lifestyles and malnutrition trigger weight gain. And the effects of the two triggers are intertwined: with super-sedentary lifestyles can we obtain enough non-calorie nutrition without exceeding energy requirements? Could compensatory appetite regulation be abolished because of certain deficiencies? Hunger is powerful enough to drive a search for food that involved processing acorns. It is powerful enough to allow food combinations to be selected that complemented each other, despite the fact that the nutrients could not be detected. Isn't it logical to assume that this same hunger could continue beyond calorific needs if other nutritional requirements have not been met?

From the beginning of time, nutrition has been a major challenge for humankind. Starting out as hunter–gatherers, malnutrition was one of the many hardships that saw longevity as what we now call middle age. Malnutrition was epidemic because of shortages – the same shortages that kept obesity at bay – and poor medical care masked the true consequences of what our ancestors ate.

In taking control of the food supply, the human race entered a new stage of nutritional history – farming. They faithfully followed the inbuilt food preferences that had allowed them to survive the Stone Age. The preferences now had a new power: before farming they were only able to select the best diet from the range of foods on offer, now they guided the range available. The projects were both short term, selecting annual crops, and long term, selectively breeding those that best suited human needs. In both plans, increasing the energy-density of food aided survival because throughout the first stage of nutritional history and well into the second, the energy balance settled body weight somewhere in the BMI underweight zone. However, as time progressed, farming allowed a new nutritional balance and the amount of meat, cereals and sugar increased at the expense of vegetables.

By the 16th century, vegetables were falsely feared, and scurvy, amongst other nutritional problems, was rife. Health suffered, but on a background of generally poor medical care, food adulteration and a life expectancy not much more than that of the Stone Agers, the consequences of poor nutrition were often overlooked. Nevertheless, the value of good food was recognised. Medicine was based around the diet and food products. When people became overweight or even obese, reducing food consumption was recognised as the only sure cure. And many of the population did experience unwanted weight gain, despite the average calorie availability being less than what scientists now believe is healthy, and despite their daily calorie needs being more than those of modern-day humans. The unbiased energy balance was tipped in favour of unhealthy weight loss, but many people became stout, particularly in what is now known as middle age, suggesting that the balance was biased towards stockpiling. Still, up to the end of the 19th century, when trim was truly in vogue amongst industrialised nations, most people managed to attain their desired physique.

The chemical revolution began in the middle of the 18th century, but was slow to have an effect. Scientists argued about nutritional recommendations because technology did not allow a categorical molecular understanding of how our bodies work. The chemical revolution made several important advances. First, it doubled life expectancy. Medicine, nutrition, energy, clean water and sanitation are central to this advance; largely the same concerns that remain today. Second, in the field of nutrition, nutrients began to be identified and a list was formed. Third, chemistry was able to make many of the items on the list, and food fortification and supplementation began. But progress has sometimes been misguided. The type of experiment and the timeframe of the studies meant that they were biased towards identifying the vital nutrients whose deficiencies are quick to show an effect. When Lind discovered the importance of vitamin C, his trials took several weeks; discovering the importance of cancer-protecting nutrients takes decades. The list does not include nutrients that offer long-term health benefits, nor does it include nutrients that improve the quality of life or increase its span. The limited scientific know-how as nutritional needs were first being defined could not identify nutrients that alter the epigenome to promote obesity in a decade, nor could it identify the impact of what parents eat on their future children's health.

Food supplements were built on a science that showed that Stone Age taste preferences coupled to food availability did not fully protect against malnutrition. Some examples of their success are impressive; but nutritional diseases have been rife throughout history. These diseases were not necessarily a consequence of food shortages: excess salt was shown to harden the pulse millennia ago and when fat became plentiful, so did heart disease. Food processing was designed to provide tasty and convenient meal solutions, and fortification replaced some of the nutrients destroyed. But health is dependent on many functional foods and non-nutrient food components; are processed foods forming too many of our meals? In May 2011 we thought that Denmark had gone mad when they stood up against the trend of fortified foods – lashing out at the yeast spread that has become a symbol of Britishness. This same yeast spread probably protected generations of the British against pellagra, but with the advent of vitamin supplements and fortification becoming a selling point for food products, the Danes became preoccupied with the question of dose and chose to err on the side of caution. The country now evaluates all fortified food before marketing. Denmark has a low incidence of obesity. Any individual fortified food is unlikely to be behind the global obesity problem, but the whole range, combined with supplements in a pill form and nutritionally poor foods rich in specific additives, results in a combination of deficiencies of some food components and excesses of others that demolish fat chemistry's protection against body fat by feeding misguided signals into the body weight regulation map.

At the beginning of the third stage of nutritional history, needs were evaluated and food production tailored to meet them; by its end, technology and medicine had revised these needs. One hundred years ago, when nutritional guidelines first began to be evaluated scientifically, the global population lacked sufficient energy for good health. Sedentary lifestyles coupled to the production of energy-dense foods have made energy insufficiency a thing of the past for many countries of the world. One hundred years ago, much of the population suffered from deficiencies of the vital amines: pellagra, scurvy and beriberi epidemics were devastating. At first, palatable foods rich in the vitamins the population lacked the most were developed. Improving palatability generally involved adding sugar, salt and fat. As time progressed, these vitamins were manufactured synthetically and packed into pills; both forms of supplementation have become integrated into many people's daily diets. Now we question the consequences of their excess: both the direct poisoning effects and the indirect health problems that result from nutritional imbalances. One hundred years ago, life expectancy was about half a century; today, for an individual of a healthy weight, it has almost doubled. Whereas our ancestors at the turn of the last century were becoming aware of nutritional problems that manifested themselves within weeks, at the turn of the millennium we are beginning to understand that nutritional problems could take decades to rear their ugly heads; some wait for the next generation.

In the fourth stage of nutritional history, one that is beginning right now, foods will be viewed with as much respect as medicine. This era evaluates the correct dose, categorising the consequence of nutritional excess with that of deficiency. It relies more on intelligence and knowledge than instinctive programming. The impact of chemical technology on food production has undermined the power of instinct in determining dietary balance, and a little bit of what we fancy may not do us any good. As for obesity, it is no longer considered a consequence of greed and laziness, but a symptom of disease. The altered fat chemistry that affects appetite and metabolism, powerfully promoting increased body weight, also triggers many other health problems. Therefore, weight regulation is no longer considered a question of vanity. Obesity is an endocrine disease: a disease that involves glands and hormone signallers. Such diseases are powerful and have far-reaching consequences. Central to the changing fat chemistry caused by obesity is the increase in hormone production as visceral fat stores swell. Despite pumping out hormones, this fat store has yet to be awarded the classification of an endocrine gland, but reaching such status is only a matter of time. It took many molecular details to launch the field of gut endocrinology, which is exclusive to the digestive tract; the fatty gut is only just catching up. And when it is considered as the fat gland, obesity will be redefined as a disease involving a swollen gland and it will be treated with the respect it deserves.

As for the cause of obesity, key consideration will be given to dietary balance, along with healthy levels of exercise and relaxation. Exercise and rest are easy for an individual to gauge, but dietary balance is a minefield, especially when the market includes products described as portions or meals that do not provide a healthy range of nutrition. Certain food components certainly contribute towards body weight gain. Sugar is one, stimulating insulin to promote body fat storage. Fat has a critical role in obesity too. High-fat diets cause changes in the way genes are used to promote obesity; no-fat diets result in essential nutrient deficiencies. In addition to quantity, there is a quality factor. Saturated fats have already been linked to heart disease. *Trans*-fats are becoming increasingly implicated in health problems. In particular, these naturally rare fats reduce PPARg signalling to promote insulin resistance, whereas more natural polyunsaturated fatty acids increase PPARg signalling potential. Protein excess has been shown, via somatostatin, to reduce the body's sensitivity to satiety signals. Many micronutrients affect appetite and metabolic rate. And then there is the effect of the balance of glutamate, phosphate, aspartate and salt on calcium balance and signalling. Excesses of these nutrients combine with calcium deficiencies and sedentary lifestyles to weaken bones, with the knock-on effect of body weight gain. Much of the food we eat looks good, tastes great and is convenient; it matches our genetically programmed taste preferences and intrinsic laziness. But the nutritional balance does not quench the appetite. Either hunger continues because of the positive feedback effect of fat and sugar, or because of a lack of nutrition or fibre in the food, or because of excesses of certain nutrients. In all cases, people eat beyond their needs.

Inheritance has often been blamed for obesity: big parents produce big children. Several twin pairs separated at birth have been shown to grow up to have similar body weights. Furthermore, twins tend to gain weight in a similar way – certainly more similar than random members of the population. At first glance it appears as though the findings suggest that body weight is determined by genetic variation. Some of these factors are powerful, others are said to have an effect only under certain environmental conditions. However, the discovery of epigenetics and, in particular, the role of high-fat diets in predestining children to obesity, reveals the importance of non-genetic inheritance: inheritance that is environmentally determined, preventable and may be reversible. Just 2% of the human DNA is in fact composed of genes; much of the rest is non-genetic DNA that allows organisms to adapt to a changing environment. It is via changes in gene use that body weight is regulated. If chromosomes have made such a high investment in such mechanisms, it seems logical to assume that their role is important.

And so it seems as though the origin of obesity is in our intelligence enabling the current food production profile. It is buried in the genes we cannot change and how these genes guide our food choices and appetite control. These taste preferences have blindly guided nutritional history, but have been unable to prevent malnutrition. And now they have a chemical weapon to ensure they get what they want, obesity is the result. The rising incidence of obesity is due to the environment we created using our intelligence to challenge Nature and feed a growing population, following our genetically programmed food preferences to guide the choices of foods made available. But we are not as clever as we thought. A lack of knowledge meant that nutritional progress was misguided. The food we have made available is not nutritionally balanced with today's lifestyle and the results of nutritional excesses and deficiencies bias the energy balance. Natural protection against unhealthy weight gain is now insufficient.

Therefore, the end of the obesity epidemic must lie in intelligence too – using intelligence to change the environment and choose what we eat with care. Intelligence caused the obesity epidemic and intelligence can cure it. Using intelligence to over-ride genetically programmed taste preferences reduced the incidence of heart disease through the first half of the 20th century. Using our intelligence to choose appropriate foods to match our individual lifestyle can reduce the incidence of obesity. And it does not take much intelligence because there is a failsafe: when in doubt, fresh is best. The foods that damage health and promote obesity are usually the highly processed options; if you are unsure of your body's tolerance of these products err on the side of caution and choose natural products. The choice is yours.

Subject Index